Fungi

Roy Watling

SMITHSONIAN BOOKS, WASHINGTON, D.C.
IN ASSOCIATION WITH THE NATURAL HISTORY MUSEUM, LONDON

Published in the United States of America by Smithsonian Books
in association with The Natural History Museum, London
Cromwell Road
London SW7 5BD
United Kingdom

Library of Congress Cataloging-in-Publication Data
Watling, Roy.
    Fungi / Roy Watling.
        p. cm.
    "In association with the Natural History Museum, London."
    Includes bibliographical references and index.
    ISBN 1-58834-082-1 (pbk. : alk. paper)
        1. Fungi. I. Natural History Museum (London, England). II. Title.
    QK603.W38 2003
    579.5—dc21   2003045659

Manufactured in Singapore, not at government expense
10 09 08 07 06 05 04 03   5 4 3 2 1

Edited by Jonathan Elphick
Designed by Mercer Design
Reproduction and printing by Craft Print, Singapore

Front cover main image: *Pholiota jahnii*; inset: *Leccinum vulpinum*; back cover and title page: *Morchella hortensis*.

Contents

Preface

The larger fungi make up for their relative unimportance numerically compared with the overwhelming number of microfungi by producing fruiting structures that are obvious and often very colourful. Although fungi were once considered to be a primitive group of plants, their physiology and biochemistry make them so different from other organisms that they are now placed in a separate kingdom. Fungi occur in almost all habitats, from luxuriant rainforests to the harsh environments of the highest mountains, the driest deserts, and even the Antarctic continent. Many of these habitats are endangered by pollution, human activity and climatic warming, and as well as acting as ameliorating agents, fungi may also play a vital role as indicators of environmental health.

The study of fungal biology, known as mycology, is entering a significant stage in its history as a whole range of scientists, environmentalists and chemists are appreciating the importance of this often neglected group of organisms. The potential rewards of understanding more about fungi are that the knowledge gained may lead to improvements in food production, discovery of new pharmaceutical products and other major technological developments.

Author

Roy Watling graduated in botany from the University of Sheffield in 1960 with a First Class Honours degree and then carried out his PhD research in mycology at the University of Edinburgh. He joined the staff of the Royal Botanic Garden, Edinburgh, where he remained until his retirement in 1998, when he was Head of the Department of Mycology and Plant Pathology. During his time there, he gained a DSc from the University of Sheffield in 1984, and in 1986 was made a Fellow of the Royal Society of Edinburgh. A former president of the British Mycological Society, the Botanical Society of Scotland and the Yorkshire Naturalists' Union, he is currently Chairman of the Scottish Branch of the Institute of Biology. Roy is also a Chartered Biologist and a Fellow of the Linnean Society, and in 1998 was awarded the Patrick Neill medal by the Royal Society of Edinburgh and in 1997 an MBE for his work both in this country and abroad. He has published several books and over 150 scientific and general papers on various mycological topics.

How important are fungi?

This question can be answered by imagining a world devoid of fungi. In such a world, the fallen trees, dead leaves, animal corpses and waste products, and all other spent organic matter would remain on the surface of the earth. The whole world would soon resemble a vast city refuse tip, and we would be literally swamped within a very short space of time. Thankfully, the invisible activities of countless fungi, aided by other decomposers such as insects, nematodes (eelworms) and bacteria, break down these dead tissues and convert them into simple compounds that can be utilized by flowering plants – plants that in turn support a whole range of herbivorous animals, themselves falling prey to carnivores. Fungi lie at the base of this food chain, of which humans are an integral and dependent part. The association of fungi with other organisms plays a vital part in many chemical pathways fundamental to life on this planet.

BELOW **Woodlands provide an enormous range of habitats for fungi, ranging from mutualistic and parasitic fungi to those which decay woody debris and litter. Here the rust-brown spored agaric,** *Pholiota aurivella,* **fruits on a fallen log.**

Fungal helpmates

Without the microscopic fungi we call yeasts, humans would not have been able to ferment sugars and other carbohydrates. We would not have been able to produce bread, a staple food of most civilizations. Nor would we have the fermented vegetable 'cheeses' (such as tempeh) in tropical countries and wine, beer and all other alcoholic drinks – a means by which early civilizations partially sterilized naturally occurring water to make it safe to drink.

And with no microfungi there would have been none of the secondary metabolites produced as a result of their activities, ranging from plant growth hormones and steroids to industrial enzymes. We would not have the whole range of antibiotics now taken for granted by the patient, at least in the developed world; or the range of food additives, such as citric acid, that are used to flavour soft drinks; or biological washing powders and other similar products that improve our living standards; or many of our familiar cheeses and, more recently, mycoproteins (such as Quorn) now so popular as meat substitutes.

More than one-third of all the food harvested throughout the world is destroyed annually by fungi and insects, often working in unison, and many of the food plants themselves suffer from diseases caused by microfungi. However, most people are unaware that such harmful fungi are outnumbered by beneficial kinds. Many of these helpful species are larger fungi.

Fungi are also involved in a whole series of important activities that biologists are only now beginning to appreciate and understand; why they carry out these processes and their long-term effects on humans are still largely unknown. Their intimate relationships with invertebrates living in the soil are important in most ecosystems, while some groups of fungi are best known for the ability of many species to form a range of beneficial associations with vascular plants in the form of mycorrhizas. In a mycorrhiza, plant roots come into very close physical contact to form what almost amounts to a single 'organism'. Residues from fungal decomposition in the world's major woodlands and forests are basic soil-components that play a vital role in long-term sustainability. The larger fungi play a major role in both the formation of mycorrhizas and the decomposition of wood or leaf-litter.

Invisible providers

Hopefully, fungi will continue to thrive – otherwise human civilization would be destroyed in a very short time, because all terrestrial ecosystems would slow down and finally cease as a result of the interruption to the flow of important plant foods. The absence of fungi would also have a major effect on the food and pharmaceutical industries. To take a single example, hundreds of thousands of pounds sterling would be lost annually if just one of the wide range of species producing simple food flavourings disappeared; industry would have to resort to the slower, more expensive method of squeezing the sap from fruit to obtain the same, naturally occurring, compounds. There are many other similar examples.

However, there is a danger that the many useful services fungi perform will remain generally unappreciated, as much of their work is carried out unseen. And despite their universal distribution, it is only when the odd mushroom, toadstool or fluffy mould appears that most people are aware such organisms exist at all. It is no exaggeration to say that the fungi form a largely invisible kingdom.

Evolution of fungi

Fungi have been recognized as fossils in rocks of most ages, dating as far back as those of the rock-like masses of cyanobacteria called stromalites that flourished on the earliest sea shores as much as a billion years ago. More recent but still very ancient fossil fungi have been found within the roots of the earliest land plants. As the vascular plants evolved, so did the fungi, and both asexual and sexual stages of fungi are recorded in coal-forming rocks and later deposits. Unfortunately, though, it is impossible to build up anything but a sketchy picture as the fragments of fossilized fungi are tantalisingly few and far between, and have not been searched for systematically.

The most spectacular example is *Prototaxites* from siliceous deposits of Old Red Sandstone nearly 400 million years ago unearthed at a famous site at Rhynie in Aberdeenshire, Scotland in the 1910s. This is thought to have been a fungus, perhaps intimately associated with an alga or a cyanobacterium to form a lichen (see p. 22). *Prototaxites* reached heights of 6.5–29.5 ft (2–9m) and was probably driven to extinction by animals grazing it, or as a result of

competition from the vascular plants that were rapidly evolving at the time. Evidence that fungi – or at least organisms resembling fungi – were widespread comes from finds in rock deposits of all ages of fossils that are likely to be their spores. To what fungus they

ABOVE **Many fungi occur in grassland, and some of these have an as yet unexplained association with certain grasses and certain mosses.**

FAR LEFT *Protomycena electra*, an agaric preserved in Dominican amber, around 15–30 million years old.

LEFT Fossil bracket fungi from the Tertiary rocks of Idaho, USA.

belonged it is impossible to say with accuracy, but educated guesses have been made, based on similarities with present-day species.

Today, the commonest group of fungi are those living in intimate association with plant cells in the roots of trees, shrubs and major cereal crops, forming what are known as endomycorrhizas (see p. 21). These are similar in all details of appearance to those found in the rhizomes (underground stems) of some of the earliest land-plants, such as *Rhynia* and *Horneophyton*, preserved as fossils in the Rhynie sandstones mentioned above. Similar structures that are even more ancient (about 450–460 million years old) have very recently been discovered, and these were probably associated with the ancestors of present-day mosses.

Evidence from fossilized amber and petrified roots dated to the time of the dinosaurs and later, suggests that as the flowering plants evolved and diversified, so did the fungi. Not only are asexual stages of fungi found, but also fruiting bodies, some of which resemble in all detail modern mushrooms and toadstools and their allies. These have been dated at 90-94 million years old, forming a real mycological 'Jurassic Park'. It appears that within all organisms there are molecular characteristics that reflect the age of their ancestors; for the fungi, this molecular clock, as it is called, fits in very well with the fossil record and indicates that the fungi diverged from common ancestors with other organisms as much as 500 million years ago.

What is a fungus?

For a long time fungi were considered to be simple plants, but they are now regarded as members of a separate kingdom, often referred to as the Fifth Kingdom. Mushrooms, toadstools, bracket fungi and elf-cups are the most obvious representatives of this kingdom, although there are very many more thousands of microscopic species.

A mixed bag

Fungi are a very diverse group of organisms drawn together by a common lifestyle coupled with their overall rather simplified structure. They are generally composed of long filaments known as hyphae (singular: hypha) which are either long tubes enveloping living matter or tubes divided by cross-walls into segments. When hyphae are grouped together they form what is called the mycelium. This makes up the vegetative, or asexual, state of the fungus.

Not only does the mycelium permeate the soil, wood or whatever substrate the fungus is using as a food base but it also allows the fungus to progress from one site to another. In this way, it resembles not the roots of a plant but a mass of the tiny root-hairs that emerge from them.

Unlike plants, fungi lack the ability to make food from sunlight and simple nutrients

LEFT **Leaves on the floor of a West African rainforest showing colonization by the vegetative stage (mycelium) of several litter-rotting fungi.**

ABOVE **The fairy club basidiolichen,** *Multiclavula vernalis,* **growing on peaty soil, Hebridean Islands, Scotland.**

by photosynthesis, although some species have evolved in unison with an alga or cyanobacterium to form lichens, and these feed on the energy-rich organic compounds manufactured by their photosynthetic partner (see p. 22).

In general, fungi do not engulf their food like most animals, but absorb it in liquid form, as a soup of nutrients. The fungi obtain these by using enzymes to break down the complex compounds made by plants and animals into simpler forms.

There is a relatively small yet important number of fungi, the yeasts (see pp. 14–15, p. 17), which in their vegetative state are reduced to chains or clusters of single-cells, and another similarly small group, the slime moulds, or myxomycetes (see p. 13), which feed in a similar way to primitive animals.

Types of reproduction

Fungi generally reproduce sexually by means of tiny, usually single-celled bodies called spores, although many fungi can also reproduce asexually by fragmentation of the mycelial stage or by the production of spores on asexual structures (specialized structures on the mycelium). Dispersal is either by spores or by the active long distance spread of the mycelium within the substrate. The number of spores produced by a single fruit-body is incredibly high. Many of them move freely around in the atmosphere and are taken in with every breath.

Depending on the species of fungus, spores are either formed within single-celled structures or attached to the outside of single or multi-celled structures. They are produced in a whole multitude of different types of fruit-bodies, ranging from totally enclosed ones or those enclosed but with an apical hole or holes, to those which look like sheets spread over the soil, wood or leaves, often taking the form of the substrate on which they are growing. Other fruit-bodies are more complex in appearance: some have a stalk, which may or may not have a protecting cap, while others resemble animal organs, such as lungs, intestines or brains. In these cases, the spore-producing cells are arranged on plates, called gills, as in the mushrooms, on projections, within pores, or on wavy irregularities.

The different types of structure are used in classification but the way the fungus disperses its spores is fundamental. Although mycologists know a little about the triggers for the development of fruiting structures and the spores on or within them, they are still a long way away from understanding the processes completely. The major factor seems

to be the availability of sufficient moisture. Once the spores are released, they may be carried on the wind or on mammals' fur, birds' feathers or the outer surfaces of small invertebrates. By such means, they can be carried a long way from the parent fungus. Some spores even enter the food tract, for which they are adapted by having thick-walled spores, and are dispersed in the dung of animals, both large and small; this has the advantage that they have a ready-made source of a rich substrate for their development.

Reproductive organs

Once a fungal spore lands in a favourable environment, it germinates to form a hypha, which soon branches to form a mycelium. If the spore's landing place is unfavourable, or the conditions change, the hypha may remain dormant for some time, maybe years. Some fungi may stay growing in the soil as mycelium for decades and fruit only rarely, if at all. They make do without sex, often propagating themselves by means of asexual propagules, in a similar way to some flowering plants and ferns, which produce asexual bulbils. By contrast, the mycelium of dung fungi, for example, quickly forms fruiting structures that last only a short time.

The hypha is the youngest stage; the mycelial stage that is formed by the branching of these hyphae may then become the dominant stage. However, fungi are usually classified on the basis of the sexual, fruiting

BELOW **Spores of mushroom germinating to form a germ-tube, hypha and hyphae and forming a hyphal network, or mycelium.**

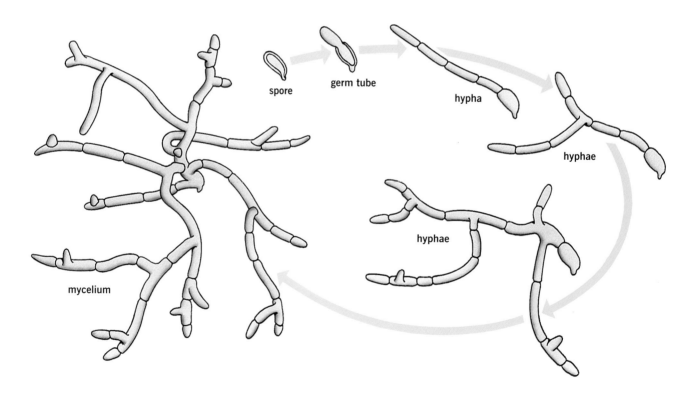

spore

germ tube

hypha

hyphae

hyphae

mycelium

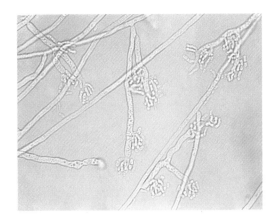

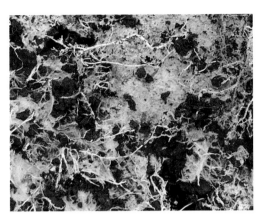

FAR LEFT **Microscopic view of a culture of the dung ink-cap, *Coprinus cinereus,* showing vegetative growth producing asexual propagules.**

LEFT **Vegetative growth (i.e. mycelium in places forming cords), of a fungus permeating and colonizing humus in soil.**

stage, with its huge range of different structures, so it is the latter which must be considered the true adult stage. Many fungi never reach the sexual stage, so fungi are very different from all other more complex organisms apart from a few exceptions, such as the plant lesser bulbous saxifrage *Saxifraga cernua,* which under certain conditions never produces flowers and reproduces only asexually by adventitious buds.

The major groups of fungi

Traditionally, the fungi were divided into five major groups, based on the structure of the reproductive organs and where the sexual spores are formed. Although this approach has changed dramatically as a result of recent research, it is still one of the most useful ways of being introduced to the hotch-potch of organisms most biologists call fungi.

Myxomycetes

This is a strange group of organisms that are often known by their popular name, slime-moulds. Although most myxomycetes are microscopic, some species are often visible, producing large, brightly coloured, slimy, naked masses of protoplasm, called plasmodia, that can move slowly over wood, moss or other surfaces. By engulfing bacteria, yeasts and decayed plant matter and feeding on them, they resemble amoebae and therefore act as if they were primitive animals, but at maturity the slime organizes itself into structures bearing masses of powdery spores similar to those found in many other microscopic fungi. However, they are not nowadays thought to be true fungi and are classified with the protists, which include a whole range of microscopic single-celled or acellular (no cells) organisms. Along with the phycomycetes, the slime-moulds have always been considered primitive.

'Phycomycetes'

Members of one of the assortment of subgroups making up this varied group of structurally simple fungi rarely produce easily visible fruit-bodies. Instead, they are generally composed of aggregates of filaments or cells, which at the most form swellings full of spores or support tiny, pin-like spore-

Simplified diagrammatic representation of the more obvious and important fungi

	Groups	Major subgroups	Latin name	Examples	Characteristics
	Myxomycetes	slime moulds	Myxomycota	flowers of tan	(i) naked amoeboid multinucleate vegetative stage; (ii) dry, powdery spores when sexually mature.
	'Phycomycetes'	Endogonaceous fungi Zygomycetes	Glomeromycota Zygomycota	arbuscular mycorrhizas pin-moulds	(i) vegetative hyphae lacking cross-walls; (ii) non-motile sexual cells.
TRUE FUNGI	Ascomycetes	Euascomycetes discomycetes pyrenomycetes Hemiascomycetes Loculoascomycetes	Ascomycota	 elf cups and morels candle snuff fungus and cramp balls brewer's and baker's yeast apple scab disease	(i) vegetative hyphae with cross-walls or hyphae replaced by budding cells; (ii) sexual spores formed in ascus in multiples of 4, generally 8. 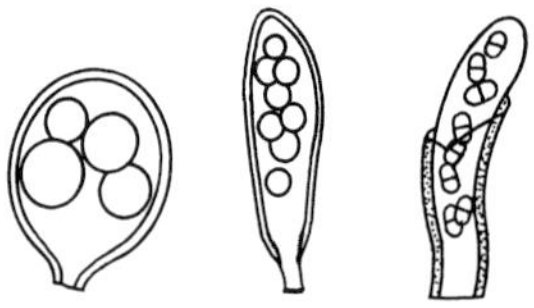 (a) (b) (c)
TRUE FUNGI	Basidiomycetes	Holobasidiomycetes 'Gasteromycetes' Heterobasidiomycetes Teliomycetes or rust fungi Ustomycetes or smut fungi	Basidiomycota	mushrooms and toadstools (agarics), bracket fungi (polypores), boletes and allies (tooth, crust and club fungi) puffballs, stinkhorns and allies (stomach fungi) Jew's ear, witches' butter (jelly fungi) brown rust of wheat, rust of hollyhocks bunk of oats, galls on maize	(i) vegetative hyphae with cross-walls, often complex, rarely hyphae replaced by budding cells; (ii) sexual spores formed outside basidium, on basidiospores, usually 4. 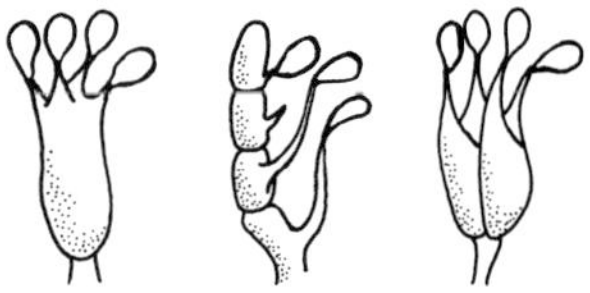 (a) (b) (c)
TRUE FUNGI	Deuteromycetes or Fungi Imperfecti	Hyphomycetes Coelomycetes	Deuteromycota	grey mould of fruit, *Penicillium* *Phomopsis* blight of juniper, anthracnose of lemon	(i) ascomycetes and basidiomycetes lacking a sexual stage, generally the former; (ii) reproduction by asexual propagules only.

producing structures elevated on stalks above the surface on which they grow and feed. The important character which unites this disparate group is that the hyphae, except those supporting the spore-producing structures, do not possess cross-walls – in contrast to the hyphae of familiar mushrooms. Because members of this group resemble algae by often forming felts on soil, plant material, food, dung, and so on, and because many species produce spores that in some stages at least have hairs for swimming (flagellae), they became known as the 'phycomycetes' or alga-like fungi. Modern studies have shown that of all the five traditional groups of fungi the phycomycetes have the greatest mix of elements, leading to further subdivision of this artificial grouping into two main groups with obvious fruiting

structures: the terrestrial Glomeromycota and the Zygomycota. The first group contains species which form symbiotic associations essential for the growth of many of our familiar grasses and herbs. The latter group, which may well be an artificial one, contains many species that live as parasites in insects and other small animals, as well as moulds that live on decaying matter. It includes the pin-moulds (for example the bread mould *Rhizopus nigricans*). It is thought that the Zygomycota are the closest to the true fungi.

Ascomycetes

The ascomycetes undergo sexual reproduction and bear their spores *inside* sac-shaped cells, called asci (singular: ascus). There are usually eight such ascospores in each ascus, in some species lined up in a row like peas in a pod.

LEFT **Two powdery mildews, the ascomycetes *Phyllactinia guttata* and *Uncinula bicornis*, on leaves of sycamore, Devon, England.**

LEFT Bracket fungus, *Pycnoporus sanguineus*, widespread in sunlit areas of tropical rainforests; Kepong, Malaysia.

LEFT Common or jewelled puffball, *Lycoperdon perlatum*, widespread in temperate woodlands.

Examples of fungi in this very large group (containing over 30,000 scientifically described species) include truffles, morels, yeasts, powdery mildews, elf cups and Dutch elm disease fungus.

Basidiomycetes

The basidiomycetes bear their spores at the top of club-shaped cells, known as basidia (singular: basidium). Sexual reproduction takes place inside each basidium, but the spores are formed *outside* the basidium on top of tiny peg-like projections at its apex. There are usually four of these basidiospores on each basidium, one at the top of each projection. Basidiomycetes include by far the greatest number of species of larger fungi, including all the familiar mushrooms and toadstools, as well as a variety of puffballs and relatives and the bracket fungi.

Deuteromycetes

The final group can almost be considered a 'dustbin' category – albeit a very convenient one, as it draws together all those fungi which do not possess a sexual stage and therefore cannot be readily identified as either an ascomycete or a basidiomycete. In contrast to most of the other groups, it does not represent a natural grouping based on evolutionary relationships.

The fungi in this group are known as deuteromycetes, or Fungi Imperfecti. The latter, alternative name is derived by analogy with the botanical term 'imperfect'; an imperfect flower is one that lacks either or both male and female sexual structures, in contrast to a typical 'perfect' flower. It is possible that through evolution the sexual cycle of deuteromycetes has been found to be totally unnecessary and subsequently completely lost, or it is very rarely formed, or it has never been recognized, even after many years of study. However, their further division into several subgroups containing a total of over 25,000 named species is still based on where and how the asexual propagules are produced – whether on the ends of hyphal filaments, in microscopic cups or flasks, or not formed at all. In the last case, the fungus reproduces and spreads solely by fragmentation of the hyphae of the vegetative stage.

The Fungi Imperfecti includes many fungi of great importance to humans, such as *Penicillium* (the source of penicillin), *Candida* and *Trichophyton* (the fungi causing the diseases thrush and athlete's foot) and various fungi in the genera *Fusarium* and *Verticillium* (responsible for wilt diseases afflicting many plants, including such important cultivated plants as bananas, flax, strawberries, tomatoes and many ornamental flowers, as well as trees and shrubs).

In some of these, the elusive sexual stages have already been found. An example is one of the most important of all fungi to humans *Penicillium chrysogenum*, the mould from which the life-saving antibiotic penicillin was isolated, which is related to the sexual stages of other *Penicillium* species, and was discovered to be an ascomycete. A much smaller number of deuteromycetes have turned out to be basidiomycetes.

Using the modern tools of the electron microscope and other techniques such as the molecular make-up of the DNA and RNA of these fungi, researchers are discovering more of the true relationships of the deuteromycetes. For example, the causal agent of frosty pod rot of cocoa (which causes millions of dollars of damage to cocoa crops annually), although forming a powdery crust to the pods is in fact most closely related to a mushroom. Originally described 70 years ago in a genus of moulds and called *Monilia rorei*, it was transferred 20 years ago to its own genus because of its structural similarities to some basidiomycetes. Recently, it has been subjected to cytological and molecular study and the ability of the vegetative stage of isolates (the first pure culture of a fungus), to fuse with each other in the laboratory has been ascertained. To the great surprise and chagrin of mycologists it is so closely related to the witches' broom pathogen *Crinipellis perniciosa* that it must be placed in the same genus. Although it may disturb mycologists to learn that the same genus can contain fungi with such different forms as a mushroom and a crust, this discovery also shows the power of the new technology available to us.

How many species?

Fungi have been generally poorly studied over the years compared with the flowering plants and many animals. There is no accurate figure for the numbers of fungal species, even for such relatively well-researched areas as the British Isles, Europe and North America. The estimate of a total of over 1,500,000 different species worldwide suggested in 1993 may be too low, as more and more studies are completed on many different groups, including palm

fungi, tropical sooty moulds, fungi associated with mosses or those associated with insects. The consensus of molecular and tropical biologists is that an increased estimate can be justified.

For instance, recent studies have found that different species of fungus, many new to science, colonize different parts of palms. Some prefer to live on the fronds, some on the stems and others on the trunk, for instance, with the exact species depending on whether the palm parts are juvenile, mature or senescent. For instance, on the South East Asian palm *Oraniopsis appendiculata*, *Arecomyces bruneiensis* occurs on the rachis (leaf stalk), *Linocarpon australiense* on their bases and *Piricanda cochinensis* only on the leaves. Researchers looking at mosses have estimated that there could be as many as 4000 different species associated with or growing on these plants. Undoubtedly, the same pattern will emerge when ferns are studied. But such figures, though impressive, pale into insignificance compared with the number of fungi living on insects. Some biologists estimate that there are as many as 10 million species of insects, and many of these have unique fungi associated with them. Sometimes, these are restricted to a precise spot on the insect, such as a segment of its leg or of the abdomen or inside its gut, e.g. *Herpomyces stylopage* on the thin-walled hairs on a cockroach's antennae. Currently, mycologists are aware only of little more than 15% of the generally estimated figure of a million or so fungi; moreover, this estimate may need to be revised upwards.

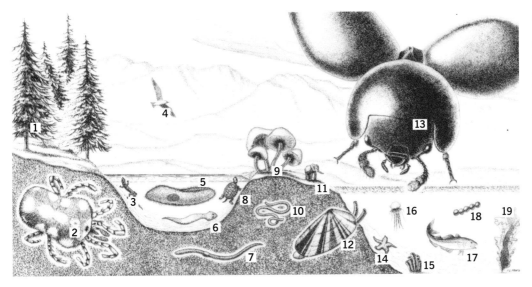

A diagram from the International Project Systematics Agenda 2000 that shows the size of an organism proportionate to the number of species in its group. More recently the fungi have been shown to be larger.

1. multicelled plants
2. spiders, mites and crustaceans
3. amphibians
4. birds
5. protozoans
6. flatworms
7. earthworms
8. reptiles
9. fungi
10. eel-worms
11. mammals
12. molluscs
13. insects
14. echinoderms
15. sponges
16. jellyfish and corals
17. fish
18. bacteria
19. seaweeds

The reasons for the lack of scientific knowledge of fungi compared with groups such as mammals, fish, birds and flowering plants originate with the early naturalists. They usually considered the fungi to be connected with the devil, and studying them at all was frowned upon by the church, right up to the 19th century, when the rest of natural history was blossoming. As a result of this taboo, scientific understanding of fungi, and especially their classification, has been hindered so much that it is no exaggeration to say that it lags almost 100 years behind that of many organisms. Thankfully, this unfortunate state of affairs is now rapidly changing as biologists appreciate the importance of these remarkable organisms and are searching for them in previously unexplored habitats.

The larger fungi

The mushrooms and their allies are regularly referred to as larger fungi (or macromycetes) and it is because of their size and often dramatic appearance that this group is the one of which humans are most aware. The larger fungi are mainly basidiomycetes but the delimitation is purely artificial. For instance, a few species of fungi easily visible to the naked eye and known as elf cups, truffles and morels are exceptions to the predominantly microscopic ascomycetes. Approximately 15,000 species of larger fungi have been described scientifically, with new ones being recognized almost daily.

Some exceptions

Lichenized species previously discussed form a large category of fungi and all the groups contain lichenized species, although the majority found in lichens are ascomycetes. Although the nearest relatives of many of the lichenized species are microfungi, lichens generally form very distinctive, easily visible structures, often covering large areas on rocks, trees or other substrates. Despite this, because of their unique status – being not individual species but miniature ecosystems, made up of a fungus living in intimate association with one or more photosynthetic partners – they are generally studied separately by specialists known as lichenologists rather than by mycologists.

Among the phycomycetes, there are a few prominent species that may form erect, wiry growths, which may reach 2 in (50 mm) or so high, but these are rare. In addition, there are a few species of myxomycete which, either in the final spore-producing stages or as they move from place to place, reach considerable sizes, sometimes covering areas the size of a large dinner plate.

Despite their size, the fungi described above, and some of the Fungi Imperfecti, which are really massed colonies of microscopic individuals, are not considered in this book as larger fungi.

Ecological divisions

Along with bacteria, protists and a variety of animals, the fungi are unseen workers that play vital roles in the food chain that starts with the production of carbohydrates by plants through photosynthesis. The contribution made by larger fungi is spread over several distinct activities, so that they can be divided into three major ecological groupings, based on the method by which they obtain their food.

Saprotrophs

The saprotrophs (from the Greek words *sapros*, 'rotten' and *trophe*, 'food') are fungi that break down the complex organic molecules in plant litter, dead woody material and animal remains, releasing and recycling nutrients into the soil in the process.

These decomposers are essential in the recycling processes that occur constantly in grasslands, woodlands and other habitats worldwide. The field mushroom, *Agaricus campestris,* is an example of a saprotroph found in grasslands grazed by cattle or sheep, where it grows on humus.

Biotrophs

The biotrophs (from the Greek *bios*, 'life' and *trophe*, 'food') are involved in intimate relationships with a wide range of trees, shrubs and smaller plants. Indeed it is generally not appreciated that over 80% of the world's flora is dependent on an association with a fungus, or in some cases several fungi even at the same time. Depending on the species, the fungus may live either in or on the root of the plant, with connections between it and the surrounding soil. Because the fungus passes on some of the nutrients it scavenges to its plant partner, this arrangement, in effect, increases the capacity

of the plant's root system to capture nutrients from much further afield, making the process much more efficient. Such associations are called mycorrhizas (from the Greek '*mykes*', fungus, and '*rhiza*' a root).

About 40% of the mushrooms and toadstools living in woods and forests worldwide are involved in such processes, and in some cases their hyphae may also help to decompose dead leaves and other ground litter. As the relationship in the larger fungi is predominantly with the outside cells of the root of the host, they are called ectomycorrhizas, (Greek *ektos*, 'outside'); another term is sheathing mycorrhizas.

By contrast, some of the jelly fungi (see p. 34) form a relationship with orchids and are known as endomycorrhizas because the fungus lives within (Greek *endo*, 'inside') the root-cells. One such fungus *Thanatephorus ochraceus*

BELOW LEFT **Wheeled marasmius, *Marasmius rotula*, here growing in Shalford, England, but which grows throughout Europe on small twigs lying on the woodland floor.**

BELOW **Ectomycorrhizal fungus of tree roots, showing the enveloping sheath and the emanating hyphae, which scavenge for nutrients.**

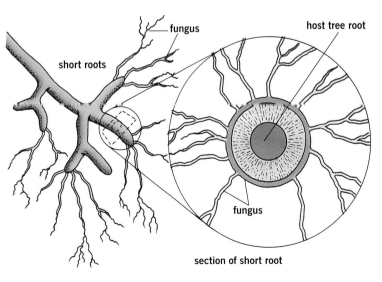

fungus

short roots

host tree root

fungus

section of short root

LEFT **A false truffle called the yellow beard truffle, *Rhizopogon lateolus*, growing in conifer woodland, England.**

is very closely related to *Thanatephorus cucumeris*, the species causing black scurf disease of potatoes and purple patch disease of cereals; while it causes disease in such cases, with orchids the relationship is very much more complex. Grasses, including all the major cereal crops, and ericaceous plants such as rhododendrons, are just two of the major plant groups that also have endomycorrhizas but in these cases, the associations are formed with microscopic species of fungi, and the myccorhizas formed in cereal crops are known as arbuscular mycorrhizas. These can be shown experimentally to scavenge for plant nutrients and transfer them through their hyphae to the host plant, benefiting both partners in the association. In the case of the orchid it is a much more complex relationship, as the nutrients are only released when the orchid actively destroys the hyphae.

A particularly close biotrophic relationship exists in the lichens, to the extent that as recently as 50 or so years ago they were considered single organisms in their own right. Lichenized fungi benefit directly from photosynthesis by forming a very close union (or symbiosis) with an alga or blue-green bacterium. In turn they benefit their partner by preventing desiccation allowing it to colonize inhospitable environments.

As well as such close relationships with plants, there are also biotrophic associations between fungi and invertebrates. In these, the invertebrate uses the fungus to help it break down resistant or complex molecules in the substrate, or it may use the vegetative state of

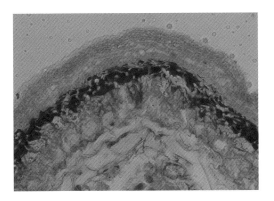

FAR LEFT **Transverse section of a root of beech _Fagus sylvatica_ viewed under the microscope showing the envelope (sheath) of fungal mycelium.**

LEFT **The agaricoid basidiolichen, _Phytoconis hudsoniana_, widespread in the mountains of the northern hemisphere.**

the fungus directly as a source of food for its larvae. In temperate countries, this kind of association is found in some wood wasps (members of the genus _Sirex_) and ambrosia beetles (in the subfamily Curculoideae); in the former, the wood wasp transports the fungus (_Amylostereum_) in special pouches. Although providing food for the wasp larvae, the fungus also often causes die-back and staining of the timber.

Probably the most publicized example of insects transmitting fungal plant disease in this way is that of the beetles, particularly _Scolytus scolytus_, associated with the spread of Dutch Elm disease, caused by a micro-fungus, _Ceratocystis ulmi_. The fungus, which is introduced by the beetle as it bores into and eats the wood, encourages the tree to plug its own water-conducting vessels, leading to wilting and finally the death of whole sections of the tree. As the fungus breaks down the wood, it makes it more nutritious for the beetles, the grubs of which also feed on the fungus.

In contrast to the cold and temperate regions, where such associations often involve microfungi, the tropics contain examples

where the fungal associates are much more obvious. The most remarkable is that with certain kinds of African and Asian termites that have evolved to depend on a group of basidiomycete mushrooms in the genus _Termitomyces_ for their food.

The termites collect plant material which is kept warm and moist in special chambers in their nests, aptly known as 'fungus gardens'. The queen inoculates this compost with fungus spores that germinate to produce mycelia, which break down the plant material; the larvae feed on the actively growing hyphae of this vegetative fungal

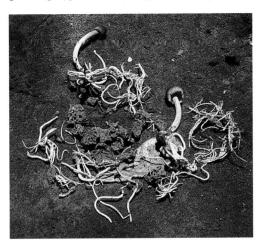

LEFT **Fruit-bodies of the termite mushroom, _Termitomyces globulus_, excavated from a termite nest and still attached to the fungus gardens cultivated by the termites; Korup Forest Reserve, Cameroon, West Africa.**

Fungal Records

Spores of fungi are all around us: we breathe them in every day – and those of larger fungi (because they rely on a liberal supply of high humidity surrounding the spore-producing cells) more particularly during the generally moister conditions at night. They can occur in astonishing numbers. In Cardiff in 1971, during the counting of fungal spores in the air, medical research calculated that there could be over 161,000 in just a single cubic metre of air. In 1987, a specimen of the giant puffball, measuring 2.64 m (8.7 ft) across, plucked from the ground before it had started to produce spores and already weighing 22 kg (48.5 lb) was found in Canada; this was estimated to have produced more than 10^{25} spores.

The vegetative stages of some fungi, too, can grow to impressive sizes. That of the 'Doncaster Monster Sandstone Fungus', found in a unique 19th century domestic dwelling carved out of sandstone in a quarry in the nineteenth century in Northern England, grew to a diameter of 4.57 m (15 ft) in just 12 months. Although originally identified as a type of tooth fungus, it was in fact probably the mycelial growth of the dry rot fungus. This notorious species is known to grow great distances through crevices between stonework and pores of bricks; one record is of a vegetative system 8.2 m (27 ft) long.

However, even such large growths as these pale into insignificance compared with the mycelium of a species of honey fungus found in the state of Michigan, in the USA. Samples of the mycelium from a wide area around were subjected to molecular analysis and found to be all exactly the same, indicating they belonged to the same individual, not just the same species. The mycelium was found to stretch for 5.6 km (3.5 miles)! Knowing the

LEFT **The dry rot fungus,** *Serpula lacrimans*, **fruiting on the inside of a garden out-building; Edinburgh, Scotland.**

LEFT **The Australian honey fungus,** *Armillaria luteobubalina*, **growing at the base of an old eucalypt; Mt. Cole, Victoria, Australia.**

amount of the mycelium found in the soil in this and other fungi and the speed by which 'fairy rings' of different fungi increase in diameter annually, it was estimated that the honey fungus weighed over 10,000 kg (10 tons) and had been stable for 1,500 years. (However, this is only half the age of some individual crustose lichens, which may be as much as 3,700 years old.)

Individual fungi such as these may occupy a larger area than any other organism. Thus in Washington State, USA, another species of honey fungus was estimated to cover 6 km^2 (2.3 miles^2). The heaviest specimen of a fungus accurately weighed to date is a sulphur polypore (one of the bracket fungi) in the New Forest, Hampshire,

England weighing 45.4 kg (100 lb). However, a specimen of the giant polypore, found in 1998 in another part of Southern England, at Kew Green, Surrey, was estimated to weigh 316 kg (697 lb), and it is still growing; at the time of publication it measures 4.9 m (16 ft) in circumference.

Even some of the common European fungi can grow to enormous sizes: a dryad's saddle 43.5 cm (17 in) across weighed just over 9 kg (20 lb), and this is an annual fungus, new fruit-bodies being formed each year, some of which may be bigger. Some of the polypores found on old trees in the tropics almost certainly weigh more.

growth. Specialized worker castes of termites maintain the preferred fungus and encourage its active growth by spraying the 'garden' with a compound which keeps the mycelia in a juvenile state. These same 'farm-workers' weed out other unwanted fungi and periodically bring the old compost to the surface so that it can be discarded. When the termites grow wings and leave the nest then the fungus can grow unchecked, and is able to continue its normal life-cycle to produce the fruiting stages – the large mushrooms that mycologists identify as *Termitomyces*. To reach the soil surface, the young, elongating mushrooms have to bore their way up from the buried 'garden' through the abandoned termite nest. Some species of this mushroom that can grow as big as a large dinner plate are collected by local people, and can feed a family for a day or so.

In the New World tropics, a parallel association is known between fungi and leaf-cutting ants of the genus *Atta*, but the fungus involved, *Attamyces*, is very different. It is an asexual fungus that produces on its hyphae small, nutrient-filled swellings, called gonglydia, on which the ants feed. In a few cases a sexual stage develops and when this happens the fruiting body that is produced looks very similar to the common European parasol mushrooms.

Necrotrophs

The necrotrophs (from the Greek *nekros*, 'corpse' and *trophe*, 'food') are the more destructive fungi that cause some cells of the host which they attack to die rapidly, leading to weakening and sometimes death, although

ABOVE **Leaf-cutter ants,** *Atta* **spp., tending their fungus gardens in Costa Rica.**

RIGHT **Rust of hollyhock** *Althaea* **caused by** *Puccinia malvacearum* **forming white, velvety coverings of basidiospores.**

in many instances the attack is not fatal. Such fungi are parasites on plants, animals or even other fungi, and include many microscopic relatives of the mushrooms and toadstools, called the rust and smut fungi. These fungi need to ensure the survival of the host so that they can attack new shoots in the future. The rust fungus generally quoted as an example

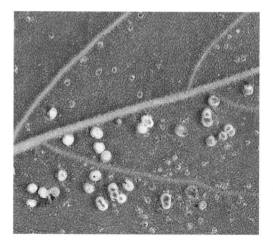

because of its economic importance is the causal organism of black rust disease of wheat but more widespread and more familiar examples are the rusts that afflict roses, zonal pelargoniums, hollyhocks, antirrhinums and other garden plants and the gall-forming rusts of acacias. Smut fungi include species causing diseases of cereal crops, such as stinking bunt of barley or a species *Ustilago maydis* that attacks maize and produces grotesquely swollen corn cobs. These are considered a delicacy and eaten in Mexico; called 'huitlacoche' or 'cuitlocoche', they have been known from Aztec times.

Other nectotrophic fungi are capable of mass destruction of cells and in such cases the fungus brings about the death of the whole

ABOVE LEFT **Galls on corn produced by the smut fungus,** *Ustilago maydis*.

ABOVE RIGHT **'Green' brooms of cacao, the result of infection by** *Crinipellis perniciosa*.

ABOVE *Crinipellis perniciosa* **fruiting on fallen liana stems; Ecuador.**

host or a particular part and thereafter feeds on the dead plant. Several larger mushrooms and toadstools fall into this category. One example of such an important pathogenic larger fungus is that which induces the formation on cocoa plants of the green leafy and monstrously swollen shoots that are commonly known as witches' brooms. The causal agent is a beautiful, pink *Crinipellis*, which is visible only once the tissue of the host dies and falls to the soil surface.

Some larger fungi, such as the common bicoloured deceiver (*Laccaria bicolor*) and the oyster mushroom, have recently been shown to use insects and eelworms as a source of nitrogen, the asexual stage penetrating the insect or worm's outer skin and digesting its

inner organs. The oyster mushrooms, which today are found in supermarkets throughout the world and which come in an array of colours, bear sticky, lassoo-like traps on their hyphae that snare unsuspecting eelworms.

There are thankfully only a few larger fungi that attack humans, and then generally only when the patient's immune system is weakened. The fungi involved are common and widespread species, such as the split-gill fungus and dung ink-cap, which can be found all over the world. They carry out their damage in the asexual stage. The split-gill fungus grows in and destroys toe-nails and soft tissue, although it has also been isolated from sputum and cerebral fluid; the dung ink-cap has been isolated from both lung and heart tissues.

Mycologists are learning more and more about other intimate and parasitic relationships between fungi and their insect counterparts all the time, and there is now a realization of just how widespread these associations are in nature.

ABOVE **The common oyster mushroom, *Pleurotus ostreatus*, the vegetative stage of which is capable of trapping and consuming eel-worms as a source of food.**

RIGHT The split gill fungus, *Schizophyllum*, here growing on wood although it is now frequently also found growing from burst bales of sillage.

Classifying fungi

The recognition of similarly structured and functioning individuals allows taxonomists (biologists who study the scientific classification of organisms) to cluster them into fundamental groups called species; because of unifying similarities, various species can be grouped into bigger units (genera; singular genus) and these into even larger groups (families) and so on, through orders, classes and phyla. Collectively, all these groups are called taxa.

However, the scientific classification of fungi has caused problems for taxonomists for generations, since fungi do not always follow what were thought to be natural rules for larger plants and animals. With the many peculiarities in their life cycles, they have been hard to accommodate into the general code of practice adopted by the international scientific community for naming organisms. Nevertheless, despite these differences, mycologists still divide the fungi into species and the other groups in the hierarchy of classification. In the wider context, the groupings of fungi into species and genera usually parallels the ecological classification outlined above and can be used to predict the ecology of newly discovered species, but this is not always the case. In some groups, such as the various species of honey-fungus (genus *Armillaria*), one species lives as a saprotroph while a very closely related species has evolved into a necrotroph.

Unlike the microfungi, the macromycetes exhibit some easily visible features that can assist in both their identification and their classification. The classification of larger fungi for a long time was simply a cataloguing of these fruiting structures, but this has been shown to be too simplified, and such an approach is now discredited. Compared, however, with flowering plants, the larger fungi have relatively few macroscopic features, and separation relies mainly on microscopic differences, such as the appearance of the spores and the cells on which they are borne.

RIGHT **The fly agaric,** *Amanita muscaria*, **a fungus which produces a white spore-print (see p. 60), when the spores are cast.**

Mushrooms

Worldwide, the most familiar larger fungi are the umbrella-shaped types known in the English-speaking world as mushrooms and toadstools, the latter word being generally reserved for species that are – or are thought to be – poisonous. Many mushrooms and toadstools belong to the major group (order) that goes by the scientific name Agaricales; these are more informally known as agarics. These have a distinct cap protecting a series of radiating plates, called gills, beneath. For this reason, agarics are traditionally referred

to as the gill-fungi. It is on these gills that the reproductive cells are produced and the spores form. Generally, but not always, there is a stem that elevates the cap and gills above the surface of the substrate.

The agarics form a relatively large group of fungi worldwide, which can be easily subdivided by differences in the colour of their spores. These may be white, cream, pinkish, brownish (ranging from bright orange-brown to dull sepia), blackish or, more rarely, a few other colours, such as greenish or bluish. Scientific classification of the agarics into different genera is based on a correlation of the colour of these spore deposits with the structure of the individual spores and whether the stem has a 'sock', called the volva, around its base, or whether there is in its place (or in addition to the volva) a ring or annulus, which may hug the stem tightly or fit loosely around it – in some cases so loosely that it may even slip off.

LEFT **The lawyer's wig or shaggy ink-cap,** *Coprinus comatus*, **which produces black spores; Ockham Common, England.**

Bracket fungi and boletes

As well as the more familiar agaric mushrooms and toadstools just described there are other, easily distinguished, groups of larger fungi. The bracket fungi, or simply 'brackets' (known in North America as shelf fungi), are more formally referred to as polypores. As the latter name suggests, they have many pores; beneath the cap, taking the place of the gills of the agarics, there is a layer of tubes in which the spores are produced. These end in round, ellipsoid or elongate openings – these are the pores, through which the spores are shed when they are ripe. The flesh of polypores is generally tough and leathery, woody or papery, and the stem may

FAR LEFT AND LEFT
The penny bun, cep or porcini, *Boletus edulis*, common in both conifer and broad-leaved forests; right, spongy undersurface of the same.

BELOW LEFT **The artist's bracket, *Ganoderma applanatum*, on beech, *Fagus sylvatica*; Wigtownshire, Scotland.**

be present or absent. In some species even the cap is lost and the layer of tubes grows directly on the substrate. Such a sheet-like structure is described as being resupinate (from the Latin word meaning 'twisted onto its back').

In one relatively small but important group of mycorrhiza-formers, the fruit-body is fleshy, as in the agarics, but also has a sponge-like layer beneath the cap; these fungi are more closely related to the agarics than the woody fungi and are called boletes.

Just as with the agarics, the classification of the polypores and boletes depends on the colour and structure of the spores, together with details of the structure of the flesh making up the fruit-body.

Toothed, crust and club fungi

Some larger fungi, both stemmed and resupinate forms, have as their spore-producing structures not pores but small outgrowths resembling teeth or spines – accounting for their popular names of toothed or hedgehog fungi.

There is a further group of fungi, known as corticioids or crusts, that lack gills, pores or teeth; in these the spores are formed on a smooth, wrinkled, folded or veined resupinate sheet. The whole fruiting body appears as if it is stretched out to form a crust over the soil, wood or substrate on which the fungus is growing; hence they are often called crust fungi.

Despite their appearance, the toothed and crust fungi are quite closely related to the poroid fungi. As with the latter, some species of toothed and corticioid fungi develop stems and sometimes caps, too.

LEFT **The brown corky spine fungus,** *Hydnellum peckii*, **characteristic especially of natural pine woods in the northern hemisphere.**

LEFT **The resupinate or crust fungus,** *Peniophora quercina*, **common on branches of deciduous oaks and beech; White Downs, England.**

If you imagine the surface of a smooth corticioid fungus wrapped around an erect cylinder, then you have an idea of the structure of another group of fungi, called the fairy clubs, club or clavarioid fungi, although sometimes these may be branched and even coral-like or cauliflower-like in appearance.

Jelly fungi

There is a small but significant group of larger fungi, called the jelly fungi (or simply 'jellies'), that differ fundamentally from the agarics, polypores and others (although they sometimes mimic them in the shapes and form of their spore-producing tissue). They differ in that the basidium is composed of not one cell but four, or in the case of some species, such as the stag's-horn fungus, it resembles a musician's tuning-fork.

Some of the jelly fungi have the cells of the basidium arranged one on top of the other, as in the Jew's ear (or Judas' ear) fungus, *Himeola auricula-judae* (also known as *Auricularia judae*) while others, such as the orange-yellow species rejoicing in the name of witches' butter, have the cells situated side by side; when the tops of the latter are viewed under a microscope their basidia resemble miniature hot-cross buns. The basidia form within a gelatinous matrix, giving the whole fungus a jelly-like feel and appearance, which accounts for the common name of the group. They can dry out and then, by reabsorbing water on the return of damp conditions, spring into life again. The jelly fungi include some very highly prized edible and medicinal fungi; also, in the group in which the basidia are longitudinally divided there are many

species that are important mycorrhizal associates of orchids.

Although in north temperate regions there are only a few species of jelly fungi that have atypical, ungelatinized, dry fruiting bodies, in the Southern Hemisphere some ungelatinized species are involved in a complex association with bugs known as scale insects in the genus

ABOVE **Jew's or Judas' ear,** *Auricularia judae*, **a common European Atlantic coast jelly fungus most frequently found on elder** (*Sambucus nigra*).

Aspidiotus, which feed on plants. The fungus envelopes an adult scale insect, and penetrates its skin to obtain nutrients as the insect continues to feed by sucking the plant's sap. In return, the fungus not only allows the insect to live and reproduce, but also offers the bug's eggs and larvae shelter from predators. The larvae can move about freely under the protective fungus canopy, although some on reaching adulthood may become infected themselves as the fungus spreads.

Mycologists know very little about the biology of these fungi but suspect that many of the temperate-zone jelly fungi (such as witches' butter) are parasitic, if not on insects, then on other fungi. In some cases, when the jelly fungus is collected, the host fungus can be seen in the middle of the jelly as a hard core. Those species of jelly fungi with basidia like tuning forks vary in appearance from gelatinous clubs or buttons to discs; they are widespread in temperate and tropical areas.

Yeasts

Yeasts are single-celled fungi that reproduce by budding. Although most yeasts – especially those used in the production of alcoholic drinks and as a raising agent in baking – belong to the great group of ascomycetes (see p. 15), there are some other naturally occurring yeasts that are stages in the development of species of jelly fungi. Some of these have lost the ability to form visible fruiting bodies and remain as yeasts throughout their life-cycle, even in some cases losing their ability to produce sexual spores. The latter are accordingly placed in the 'ragbag' group of deuteromycetes or Fungi Imperfecti (see p. 17).

LEFT **The stag's horn,** *Calocera viscosa*, **a common north temperate jelly fungus on conifer wood; Tømte, Norway.**

Puffballs, stinkhorns and relatives

As well as those fungi that eject their spores, several groups of fungi, such as the puffballs and stinkhorns and their relatives described below, which also include the famously edible truffles, do not have this capability. Instead, the spores break free to take their chance in air currents, are released by physical impact from a passing animal or high winds to be carried off by them, or are distributed by insects.

Fungi that produce their spores inside the fruiting body belong to the group known as the stomach fungi, or gasteromycetes. The most familiar of these to most people are the puffballs. These have a dry, globe-shaped or stalked thin-walled sac containing dry spores. The flexibility of the outer layer of the sac helps to release the spores. Some giant puffballs, *Calvatia gigantea*, can attain diameters of 1 m (3.3 ft) or more; the record is held by one found in New York State, USA, in 1877, which measured 1.62 m (5.3 ft) across and was from a distance mistaken for a sheep.

Another subgroup of gasteromycetes have spore sacs with tougher skins that lie on the soil and resemble small to medium-sized potatoes; these are known as the earthballs. The name earth stars aptly describes the

RIGHT **The giant puffball, *Calvatia gigantea*, widespread in grass along woodland margins.**

FAR RIGHT **The earthstar, *Geastrum fornicatum*, found most commonly in beech woods.**

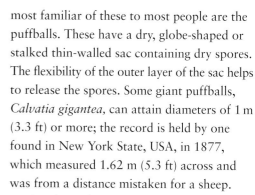

shape of a third group's fruiting bodies at maturity, when the outer of their two skins rolls back to resemble a star with the spore sac in the centre. In some species, the split skin is bright red, orange or yellow. Another group of gasteromycetes, found mainly in the Southern Hemisphere, have soft, often brightly coloured and pouch-like fruiting bodies, and are called tobacco pouches. Their spores are probably dispersed by slugs and snails.

Also included among the stomach fungi is a group called the bird's-nest fungi. In these, the fruiting body when it opens at maturity looks like a small nest; the spores develop within it in small rounded packets resembling the eggs of a miniature bird. Some species have evolved a 'spring' that helps to shoot the spore-packets for several metres away from the 'nest'.

LEFT **The white egg bird's nest fungus, *Crucibulum laeve*, widespread in many parts of the world growing on old, tough herbaceous material and woody debris.**

LEFT **The common earthball, *Scleroderma citrinum*, common in many parts of the world but always associated with trees.**

Some stomach fungi have strong, distinctive smells to attract insects or other invertebrates; so intense are the odours of a ripe fruiting body that they can travel impressive distances. One group resembles the earthballs in that their fruiting structures are about the same size and the fruiting body is totally or partially buried in the soil. As they can be confused with the edible truffles, which are ascomycetes, they are called false truffles.

The stinkhorns have evolved fruiting bodies with bizarre, often strikingly coloured shapes, referred to as phalloid, with a long cylindrical spike capped by a darker head, bearing the spores. The spike erupts from a partly buried, egg-like structure. In many of the tropical species, the spike is surrounded by a big, lacy, 'skirt'. Another unifying feature of this group is their habit of producing their spores in a foul-smelling mucus, with the odour of rotting flesh. This attracts flies and other carrion insects, which feed on the mucus. Like bees gathering pollen from a flower, they become dusted with spores, and carry them away with them when they have finished feeding, in the process helping to disperse the fungus. This group of fungi seems to have evolved to a peak of ingenious design in Australasia from species which are red, orange, green or blue to species which have star-shaped fruiting bodies, and lantern-like fruiting bodies, to colourful tongue-like structures and cages. The reasons for this are not clear, although some mycologists have hypothesized that it is linked with the long isolation Australia has experienced over geological time.

Larger cup fungi and morels

A relatively small group of larger fungi are ascomycetes, in which the spores are formed in microscopic sacs. Among this group are the subterranean true truffles and the more frequently seen elf-cups, which may reach several centimetres in diameter and, like many basidiomycetes, may have their cups elevated above the ground on stalks. In some species, the cup may become wrinkled or wavy and folded, and such species with stalks are known as saddle-fungi, brain fungi or lorels, *Gyromitra esculenta* (also called lorchels or

ABOVE **A large tropical stinkhorn, *Dictyophoa multicolor*, plucked when young from a fallen tree in the rainforest, and incubated in the laboratory; Pasoh Forest Reserve, Malaysia.**

RIGHT **The scarlet elf cup,** *Sarcoscypha austriaca,* **found on twiggy litter amongst leaves or lying on soil often appearing very early in the year.**

false morels). Unlike the gills of an agaric, which are protected by the cap, the reproductive layers in these ascomycete fungi are unprotected from the elements. Despite this, the spores themselves are protected within the sac before being forcibly ejected when mature – often in a visible cloud of dust-like particles which can be seen if the fruiting body is held up against the light.

In a different group of large ascomycetes, called morels, many cups are aggregated onto a stalk to form a honeycomb-like structure. Others form galls on trees and these resemble strawberries – indeed they are often called aboriginal's strawberries. Both the morels and the aboriginal's strawberries are great delicacies; by contrast, some of the saddle and brain fungi can cause human fatalities.

Although other ascomycetes may form easily visible fruit-bodies, these are usually aggregations of microscopic fruit-bodies that may or may not be seated in or on a special quite prominent stroma (a prominent cushion or rough mass of vegetative tissue). One example of the latter is a species that grows on the dead wood of ash trees and goes by the odd names of cramp-balls (from an old belief that carrying them around would ward off cramp) or King Alfred's cakes *Daldinia concentrica* (from their resemblance to the cakes said to have been burnt by the English King Alfred); others are the equally colourfully named candle snuff fungus, *Xylaria hypoxylon* and vegetable caterpillars. Although regularly collected by naturalists, these are not generally considered by mycologists to be larger fungi.

RIGHT **The morel,** ***Morchella hortensis*, in a newly made flowerbed growing on wood mulch and pushing up between decorative stones.**

LEFT Cramp balls, or King Alfred's cakes, *Daldinia concentrica*, grows on old wood and most frequently in Britain on ash, *Fraxinus excelsior*.

BELOW LEFT The candle snuff fungus, *Xylaria hypoxylon*, very common on old wood, mossy stumps, and logs.

Evolution of new species

Fungi have long been used as laboratory tools in population studies, genetic experiments and theoretical concepts on speciation (the process by which new species are formed over time in the course of evolution). However, little is really known as to how the results obtained from such research relates to what happens in nature.

Close relationships

Ectomycorrhizal fungi are considered to have evolved closely in parallel with their host trees; the relationship may be quite narrow, with a fungal species or group of species closely associated with a particular group of trees. In contrast a fungal genus may be linked only to one or two genera or a single family of trees; in other cases the fungus may have a very wide spectrum of hosts. This seems to indicate that isolation with the host has led to speciation of the fungi but, in contrast, there is evidence that some fungi have switched their host tree and migrated along with the host plant to new sites. This may explain the ambiguity of the presence of some ectomycorrhizal fungi in many, often climatically different, areas of the world; now it is necessary to see whether they are genetically the same. Both ectomycorrhizas and endomycorrhizas, and those involving both microfungi and macrofungi, have evolved many times over many years, some being many millions of years old and dating back to the time that the land was first colonized by plants 400 million years ago.

It appears that the close association of mushrooms in the genus *Termitomyces* with various genera of termites and their counterparts among the leaf-cutting (attine) ants has developed independently in the Old World and New World tropics respectively. It also seems from the results of molecular research and the studies of the intimate relationships involved that this has occurred over a long period of time.

This is paralleled in the lichenized fungi, which have undoubtedly evolved several times because the relationship is recognized in a whole series of unrelated fungi, including agaricoid, clavarioid and crustose basidiomycetes and also ascomycetes. However, even though several evolutionary lines can be identified in the latter, the most widespread and commonly seen species – such

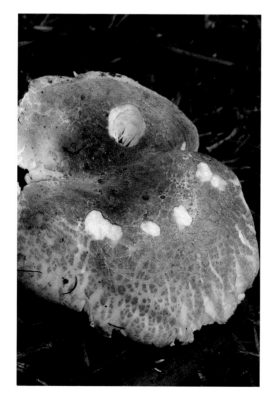

RIGHT **The green cracking russule,** *Russula virescens*, **always associated with broad-leaved trees in many areas of the world. .**

Alternative classifications

There are several exceptions to the broad outline of different groups of larger fungi given above, and recent researches have shown that this generally adopted classification is very simplistic, although useful. If microscopic, developmental and anatomical characters are taken into account, then what are likely to be much more natural groupings can be assembled. Some of these groupings were suggested by mycologists as a result of field studies many years ago – in some cases based on little or nothing more than a hunch. Currently, some of these hypothetical groupings are being confirmed by molecular studies. Thus fungi with spiny, dark coloured spores are found among the resupinate, poroid, hydnoid and possibly agaricoid fungi, and a grouping based on this character is supported by the production of chemicals which are exactly the same, similar or can be linked by well-known chemical reactions in other species. Similar examples have been found to be widespread among the larger fungi, and even link those fungi which eject their spores forcibly with those that cannot eject their spores – for example, linking the false truffles and earthballs with the boletes.

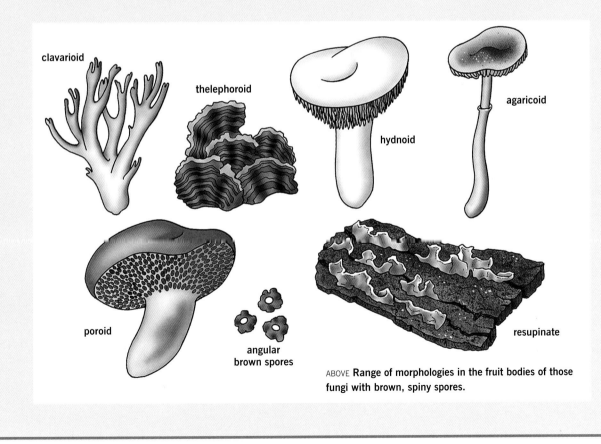

clavarioid

thelephoroid

hydnoid

agaricoid

poroid

angular
brown spores

resupinate

ABOVE **Range of morphologies in the fruit bodies of those fungi with brown, spiny spores.**

as reindeer moss and old man's beard lichen – have a common ancestor among the microscopic ascomycetes. Again, molecular studies have confirmed these observations based on appearance.

There are generally few parasites among the larger fungi, and those that do occur are necrotrophs (see p. 26). Special relationships are found between the disease-causing basidiomycetes known as rust fungi and smut fungi and their host plants, in which the life-cycle of the fungus is closely linked to the life-cycle of the plant – and in some cases two plants, because different stages of the same fungus are found on different, unrelated, hosts. Such intimate relationships are not found among the agarics and their relatives. However, the close relationships observed in the rust-fungi and smut-fungi indicates that they have undergone a long period of evolution, which may not be the case with the agarics and relatives.

Many of the ideas put forward concerning the speciation of fungi are purely hypothetical. These include enclosure of the fruiting body to withstand arid conditions, and developing a strong odour when subterranean to encourage finding by mammals and by so doing help in distributing the spores by transport in their guts; often, as with the truffles the odours are targeted preferentially to a particular group of mammals. Other examples are seen in the stinkhorns that evolved their foetid smells to attract carrion insects; hairy surfaces to the caps of fungi occurring on wood to withstand alternating dry and wet conditions; resupinate fungi growing on the undersurface of wood to enjoy the constant damp microclimate above the soil surface; puffballs and earthstars growing at the edge of the canopy to receive drips of water from the leaves that are heavier

BELOW **The fairy club,** *Clavaria argillacea,* **widespread on peaty soils, sometimes fruiting amongst potted plants.**

than falling rain and thus able to trigger spore dispersal. The list is extensive but mycologists urgently need to carry out more controlled experiments to confirm or refute such theories, which are based on field work.

Competition

Competition is probably the major driving force for speciation in fungi, and this aspect has been examined experimentally in the laboratory. The application of a concept called the r/K strategy (a formula used to explain how competition amongst vascular plants results in the mosaic of individuals found in a plant community) has been applied with reasonable success to fungi. Both 'r' types, opportunist species that succeed best in variable or unpredictable environments, reproducing rapidly, and 'K' types, highly competitive species that do better in stable environments where many other species are established and that do not need to reproduce so quickly, can be recognized among communities of fungi. The adoption of one or other life-strategy allows a fungus to exploit its chosen nutrient resource, a food which may be less favourable to competing fungal species. This is a mechanism by which the evolution of the fungus is driven, leading to greater efficiency and finally more specialized, dominant, and unique habits.

However, fungi are rather different from other organisms in that they are frequently able to change life-strategies and lifestyles, so giving them more than one opportunity during the colonization phase. This remarkable plasticity, and their ability to hold on to many hidden traits, which, though quite major, can give no hint of their presence until they are suddenly expressed as easily as a light being switched on, has evidently been of great help to the fungi in their evolution.

When and where?

Fungal fruiting is erratic and unpredictable. Whereas some fungi may fruit annually, others appear only after very much longer intervals.

When to look

In western Europe, naturalists interested in larger fungi start to prepare for their collecting season in August and continue their activities until mid-October. However, many fungi start fruiting earlier than August (for example chanterelles, *Cantharellus cibarius*, appearing from May in the border country of Scotland), while others continue well into the late autumn, their persistence depending on the arrival of the first hard frost. Several larger fungi continue to fruit into November or even December in warmer areas of Europe that are unaffected by frost and sometimes go on to produce a second flush of fruiting bodies. A few species may also be found in April, although some, such as the St George's mushroom, lorels and morels, are found only in the early part of the year.

In arctic regions the fruiting season is telescoped into a very brief period between the time when the snows melt and the permafrost relaxes its grip and the return of harsh conditions, and the situation is similar on high mountains. Nevertheless, the majority of species found in these environments are similar to those of more equable climates. In the Rockies of North America there is a small group of fungi that fruit at the edge of melting snow banks (giving them the common name of snowbank mushrooms). Some even live under the snow, their fruiting bodies pushing up through the white blanket when they are mature.

In areas with a Mediterranean climate the spring is often a productive season although fruiting may come to an abrupt end as the drying effects of the summer take hold, recommencing as soon as the rains come.

In North America, good collecting seasons are generally the same as those in Europe, with more prolonged periods as one travels from Mediterranean and rain-shadow communities to the hot, humid tropical areas of the Gulf States; along the western seaboard of the United States good collections can be made

BELOW **The autumn or winter chanterelle,** *Cantharellus tubiformis*, **widespread in the pine woods of the northern hemisphere. It is edible but not as sought after as** *C. cibarius*.

RIGHT **The snowbank mushroom,** *Hygrophorus goetzii,* **which fruits on the margins of conifer forests in melting snow banks; Three Sisters Wilderness park, Rockies, USA.**

even around Christmas and the New Year.

In many northern parts of the world September is the height of the fungus season, but if one wishes to find the full range of species then collecting all the year around is necessary. There are many fungi, particularly resupinates growing on wood, which are at their best for identification during the winter months. Various small agarics can also be found at this time of the year.

In the rainforests of West Africa and South-east Asia, although there are peaks in the production of fruiting bodies, good collections can be made nearly all year round. In more seasonal tropical regions, however, fruiting is more dependent on the onset of a rainy period; in some regions of West Africa, for example, many species can be found in the early parts of the year. In Malaysia and Thailand the optimum collecting period occurs a little later, often with a repeat of the same species in August and September.

In the southern parts of South America, and in Australia the seasons and collecting times are the reverse of those in the temperate regions of the northern hemisphere, for instance, in February and March instead of August and September.

Where to look

Fungi have colonized all the major habitats of the world. They occur in marine and freshwater environments, high mountains and lowlands, permanently waterlogged swamps and harsh deserts, all types of woodlands, including rainforests, deciduous woodlands, coniferous forests and plantations, and grasslands, from savannas and steppes to meadows and pastures. They have even invaded the human world, where microscopic forms are able to grow in extreme conditions, such as on printed circuits in computers, damp camera lenses, and on stored foodstuffs in cans and freezers.

Fungi are such adaptable organisms because of their simple vegetative structure, which can spread easily and extensively given a suitable food source, coupled with their rather simple metabolism, which can be shut down, enabling them to remain dormant (often for very long periods of time) until favourable conditions return. Fungi also do not need to reproduce sexually on a regular basis, if at all, but can colonize and remain active in sites where there is no evidence of fruiting structures.

The larger fungi are generally confined to the wooded and grassland communities, from mountain tops, where they can be closely associated with dwarf willows, to valley bottoms. Lowland woodlands resemble rainforest communities inasmuch as they are very productive. Many easily seen fungi can be found in more arid localities, such as dwarf willow communities in the hollows (slacks) between sand dunes in temperate and boreal zones.

Woods and forests

From the great rainforest regions of South America, Africa and South-east Asia and the eucalypt woodlands of Australia to the drought-resistant sclerophyll forests and broad-leaved woodlands of warm and temperate areas, and the great coniferous forests (or taiga) of the far north, wooded habitats worldwide all support a high diversity of larger fungi.

Such habitats contain a great variety of trees, shrubs and other higher plants, providing fungi with a wide range of substrates, such as leaves, twigs, branches, trunks and stumps, all in various stages of decay. Other fungi are intimately associated with tree roots in myccorhizas, or simply gain their nutrients from the humus; such saprotrophs are called humicolous fungi.

Within a forest, there are many different sites with their own microclimates, resulting from differences in the surrounding vegetation, and each favours a particular group of fungal species. Closed woodlands where all the available niches have been filled, wherever they are in the world, are rich in fungi.

The greater number of higher plants present in a locality, the greater the diversity of fungi, both macromycetes and microfungi. It is hardly surprising then that with their great diversity of plants, the rainforests support a huge number of species of fungi. However, despite the fact that tropical forests are rich in species they are, surprisingly, not necessarily dramatically richer than a similar area of ancient woodland in Europe or North America or along the backbone of the Andes in South America – in contrast to the situation with many other groups of organisms, such as bats, birds or ants.

Woodland communities are among the best and most rewarding collecting grounds as they often have such a diverse group of fungal inhabitants, each species growing in a particular ecological niche.

The nutrients available to fungi in a woodland may vary with the time of year, and may occur in major flushes rather than a steady flow. These irregular patterns may be on a small scale, as when a heavy shower of rain liberates organic materials bound to clay minerals in the soil, or on a large scale, as in the rhythmic release of different chemicals by a tree passing from full leaf to leaf fall and

LEFT **The montane russule,** *Russula alpina*, **associated with both lesser willow,** *Salix herbacea* **and Alpine bistort,** *Persicaria vivipara* **in the Scottish mountains.**

finally slowing down its vital activities in preparation for colder or markedly drier periods. Sometimes, the change is more catastrophic, as when a tree crashes to the ground after a storm, altering not only the flow of nutrients but also the amount of available sunlight and humidity.

Fungi found on the forest floor have different lifestyles and needs, and various species prefer different habitats and substrates – this is an important factor in understanding woodland diversity and its conservation. For example, some of the fungi growing on dead wood fruit very quickly on newly-cast branches, while others colonize only very rotten, sodden wood; some species prefer branches or small twigs on which to fruit, while others require large boughs or stumps.

Each region has its own very specialized, rare types of woodland, containing its own scarce fungi; a good example is the Atlantic shrub with hazel and oak found in the western British Isles, with fungi such as hazel gloves and glue fungus, *Hymenochaete corrugata*. The latter species is so-named because it can prevent the dead twigs on which it lives from falling to the ground by glueing them together in the canopy. This helps the glue fungus avoid the competition for its site and food supply that would occur if the twigs were allowed to reach the ground, where they would be available for other fungi to use. Many such unique localities, providing a refuge for rare fungal species, doubtless remain to be discovered.

In the rainforests of the world, with their very high tree canopies, various species of agaric intercept twigs and leaves and bind them together. The mycelia of these fungi form hammocks above shoulder height and capture woody debris as it falls, utilizing it before it reaches the soil surface where it would be colonized by competing soil fungi – a remarkable and very successful strategy.

Plantation woodlands of non-native trees are a common feature in many parts of the world and in all major climatic zones. These can be fruitful collecting grounds and may be home for some unusual species that may have been introduced along with the plantation trees. In Europe and South America, for instance, there are species of North American fungi that were undoubtedly introduced from their native home with their hosts, for example the slippery jacks *Suillus placidus*, which was introduced into Europe with the Weymouth pine *Pinus strobus*, and *Suillus cothurnatus* from southern USA brought into South America with *Pinus elliottii*. Similarly, in Africa and Central and South America fungi from Australia have been inadvertently introduced with the eucalyptus trees planted there, for instance the devil's foot fungus *Pisolithus* spp., and the small cup fungus *Zoellneria eucalypti*.

However, native woodlands appear to be the most productive for fungi and, in general, the older the woodland community the larger the number of species of fungi it contains. Surprisingly, the basic families of larger fungi, especially the agarics, are found in both tropical and temperate forests, although there are slight differences in the abundance of members of particular genera. However, a mycologist from the temperate zone would be quite at home with the major groups found in the tropics and vice versa.

The stipitate polypore, *Microporus xanthopus*, one of the most widespread tropical fungi growing on twigs and branches; Lam Ru National Park, Thailand.

BELOW **A tropical elf cup, *Cookeina sulcipes*, characteristic of rainforest communities as far apart as South America, West Africa and Southeast Asia; Jatun Sacha Bio Reserve, near Tena, Ecuador.**

The rainforests of the Old World, such as those in Africa, Malaysia and Thailand, are chiefly ectomycorrhizal, so genera of fungi familiar to European and North American mycologists are consistently found. In the Amazon basin, on the other hand, the rainforest is predominantly endomycorrhizal, so there is a distinct lack of ectomycorrhizal fungi, such as boleti, russules and milk-caps. However, in both these kinds of tropical forest the fungi that decompose wood and ground litter are much the same as those in the northern regions of the world. In this respect, the larger fungi differ less between the major regions of the world than the flowering plants.

There are, nonetheless, some notable exceptions; for instance the genus *Cortinarius*

is exceedingly diverse in the north, where a woodland may contain many hundred species, but in the tropics of South-east Asia it is a great surprise to see a single member of this genus. Similarly, boleti in Australasia are rather distinctive and the genera found there are either absent, rare or represented by only a limited number of species in comparison with the Northern hemisphere.

Grasslands

Old, untreated grasslands in most parts of Europe are rich in larger fungi – especially a group of agarics with pink spores, a range of brightly coloured agarics in the genus *Hygrocybe*, called wax-caps, several fairy club-fungi, and the similarly shaped but unrelated earth-tongues (also called snake tongues). Some of the fruiting bodies of these

LEFT **The aboriginal paintbrush, *Podaxis pistillaris*, probably a complex of very closely related species found in all the major desert areas of the world; Namib Desert, South Africa.**

fungi form large intermixed 'troops', while others grow in 'fairy rings'. Many of the same species occur on lawns, especially those with a long and uninterrupted history of management. Compared with the fungi of the temperate grasslands, few details are known of those inhabiting the huge areas of grassland in the warmer parts of the world.

Even though they may have been regularly burnt over many generations, some of the major non-temperate grasslands of the world still support the fruiting of larger fungi, including some rather specialized species. The steppes, for instance, are characterized by members of the puffball group, with many bizarre species recorded. Deserts and semi-deserts, too, contain unusual relatives of the puffball group, as well as specialized agarics, in which the cap remains tightly clasping the stem as it grows, and the spores are dispersed by abrasion of the tissues by sand blasts. In such arid regions (as with the sudden blossoming of flowering plants) there may be mass fruitings of fungi, especially after a passing flash flood.

Among the major features of the grasslands of Africa – and previously of North America, before the buffalo of the Great Plains were almost exterminated – are the huge herds of herbivores. Herbivore dung is an excellent example of a food resource which is ephemeral, as the wild animals make seasonal movements, or even if constantly present in the environment (as when cattle and other domesticated herbivores take the place of the wild ones) is not always found in the same place. The first requirement for the specialized group of fungi that have become

LEFT **A dung agaric,** *Psathyrella coprobia*, **widespread especially on droppings of domestic animals in the Northern hemisphere; Kindrogan, Scotland.**

adapted to living only on dung is that they have to be able to locate their often unpredictable food supply. The spores of the fungi either lie dormant in the soil until they come into contact with the dung or adhere to plant food eaten by the herbivore and pass unharmed through its gut, so that they can emerge with the faeces and immediately start to grow.

Although grassland and savanna communities are especially rich in dung fungi, they can be found from the coldest regions to the tropics, wherever land animals live, in a whole range of habitats, including woodland, their numbers and variety depending on the richness of the fauna.

Moorlands and heathlands

The moorlands and heathlands that are so characteristic of regions with cold, damp climatic conditions generally contain relatively few species of fungi. Nevertheless, they can be rewarding habitats in which to collect larger fungi; this is especially true of many of the heathlands at higher altitudes in the colder parts of both Northern and Southern hemispheres, where there are mixtures of arctic or antarctic plants and alpine (mountain) plants. These plant communities can be found fringing the Arctic Circle, on high plateaus such as Tibet and on the South Atlantic Islands, for instance. Also, in tropical areas similar communities can be found at the tops of mountains such as those in Malaysia. They are all characterized by soils that are rather acidic and contain large amounts of humus.

It is only when there are higher plants, such as those that form ectomycorrhizas, which may form prostrate woods of great extent – as on the Arctic islands such as those of the Svalbard group – that there is an increase in the variety of larger fungi. Even so, there are never great quantities of woody debris in such areas and what there is rarely includes large trunks or branches. Both these factors reduce the number of ecological niches which are present – in contrast to a forest including tall, mature trees, which includes many massive fallen trees and parts of trees.

Moss beds

Mosses (and to a lesser extent liverworts) make up important plant communities in a great range of moist habitats, from the arctic

and antarctic wastes to large monocultures in the rainforests. These plants form specialized communities on banks or on the ground under trees, or even among grasses in damp pastures. These generally have distinct microclimates, with a slightly higher temperature and lower moisture loss than the

ABOVE **A sphagnum bog agaric, *Galerina paludosa*, widespread in the northern uplands and montane areas of Europe.**

surrounding vegetation; these differences are of great benefit to fungi growing among mosses on exposed rock boulders and outside the protection of a woodland. The moss helps to protect the fungus from drying out and the slight rise in temperature probably encourages its vegetative growth. Sphagnums are specialized mosses of wet, acidic habitats, and sphagnum bogs also contain their own specialized fungal inhabitants.

Recently, mycologists have been investigating links between larger fungi (especially elf-cups and some smaller agarics) and the mosses and liverworts. It is surprising how little is known of the intimacy of such associations; future work will undoubtedly reveal fascinating details.

Garden dwellers

Any habitat, even a garden in the centre of the city, has its own assortment of fungi, sometimes including rare species. Some have undoubtedly been introduced from other habitats or even other countries, but there are several widespread woodland and grassland fungi that have adapted themselves quite successfully to life in the human environment. One example is the stalked puffball. This normally lives in sand-dunes around the Atlantic coast of Britain, but after the Second World War was found growing on bomb-damaged buildings in London.

Apparently many species have modified their fruiting patterns significantly and can now be found in gardens. Various fungi are very resilient and can colonize the most unexpected places in the garden. Some are even able to grow beneath tarmac, the

fruiting bodies pushing up through and fracturing the surface, or beneath paving slabs, fruiting in the gaps between them.

The increased interest in gardening in the West has involved major changes in plant husbandry and this has had a dramatic effect on the range of larger fungi now occurring in gardens. The widespread use of bark-chips as a mulch has enabled several fungi, which otherwise would have been restricted to the ground-litter or on small fragments of wood in conifer woods, to spread rapidly in gardens. In some temperate regions of the world, the

BELOW **The winter stalked puffball,** *Tulostoma brumale*, **found on British sand-dunes, especially those bordering the North Sea.**

species involved include hallucinogenic fungi, and on some occasions also highly toxic species. In such cases, the fungi have become a potentially serious public health problem.

Introduced fungi

There have been several records over the years, particularly in the southern parts of England, of a range of bizarrely shaped Australian stinkhorns and puffballs that have grown from spores that have apparently been introduced in packing, in soil and attached to plant material. Some of these alien fungi die out, but others gain a foothold in the new country, spreading from the original place of introduction.

Wherever people have introduced plants, especially trees, into areas outside their native range, it is certain that fungi have accompanied them, either as intimate mycorrhizal associates of the roots or as 'hitch-hikers' on leaves or bark. Examples include the slippery jacks and the Australian false truffle *Hydnangium* growing with eucalypts in the Falkland Islands and Europe. Another example is a relative of the truffles, *Paurocotylis pila*, that was was first discovered in the British Isles in Northamptonshire in 1954; probably originating in New Zealand,

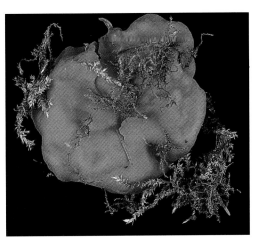

it has been recently seen in several sites in the Lothian region of Scotland as well as on the Orkney islands.

Some of these fungi colonized new areas a long time ago – doubtless not long after the first explorers left the shores of Europe and others returned from their travels. One of the most conspicuous of these early introductions was *Leucocoprinus birnbaumii*. First found in the Prague Botanic Garden in the early part of the 19th century, it soon appeared in similar places throughout Europe, and is nowadays even more widely distributed because it often turns up in potted plants purchased from nurseries and supermarkets.

Unwelcome visitors

The most notorious accidentally introduced larger fungus is, without question, the dry-rot fungus (see pp. 68–69); this species alone causes many millions of pounds' worth of damage to domestic property.

Its original home is in the forests of the Himalayas but it is now quite at home with humans in most temperate areas of the world. Mycologists do not know how the dry-rot fungus originally arrived in the West, where it has become so literally at home – it may have moved slowly from civilization to civilization, as if island-hopping, or perhaps it was suddenly imported from Asia. What is certain is it that it made its way on infected timber. Its spread in certain European countries is fairly well documented, as it moved from town to town through the use of infected timbers by traders.

RIGHT **The yellow parasol, *Leucocoprinus birnbaumii*, occurs in greenhouses and is even found amongst potted plants in homes and offices. It probably originated in Australasia.**

Collecting and studying fungi

The collecting of larger fungi, unless for the pursuit for food, should not be carried out in a haphazard way if one wishes to understand and fully appreciate one's finds. Careless collecting will result in poor material that is useless for later examination, identification and preservation.

Careful collecting and effective preservation are very important, because only a small number of species are known worldwide and there is a distinct possibility that an enthusiastic amateur may find a new or rare species. If so, the specimen needs to be kept safely and transferred to a national collection for subsequent examination by an expert who can confirm the identity of the find.

It is important to remember whenever possible to collect material in all stages of development. There is very little evidence that this will damage the overall population of the fungus, even if – as is often the case – a solitary specimen is picked; a single specimen is valuable if that is all that is available. A few good specimens, which can be well documented, are of far greater worth than a whole range of poorly collected and damaged ones. The beginner's greatest failing is to collect too many specimens on one occasion and then not have enough time to make the necessary notes. This also can apply to

LEFT **Collecting equipment should include: containers in which to place specimens, a knife, a small saw to procure the specimen complete, waxed paper to wrap-up bulky material and a tape recorder to make field notes. A truffle rake is optional.**

experienced, but undisciplined, professional mycologists when they visit a new and unfamiliar location for the first time; they too may be tempted to over-collect, to the detriment of their results – making any expedition or collecting trip ineffective and expensive.

Several distinctive larger fungi can be identified in the field without the need for collection, but for identifying many species and for carrying out more detailed study specimens must be picked for later inspection, including examination of distinctive microscopic features. Currently, there is no indication that such collecting threatens the existence of fungi, just as picking the fruit from trees does not damage the trees. Instead, the single most important cause of damage when gathering fungi is trampling, which leads to the collapsing of the air-spaces in the soil and changes in water patterns that can destroy their subterranean mycelia.

How to remove fungi

When it is essential, carefully dig up or cut out from the wood the *entire* fungus, handling it as little as possible. Try not to finger the stem more than necessary, to avoid destroying distinctive structures. There are few enough characters to work with when fresh, so obliterating or damaging them as a result of poor collecting techniques wastes time and can lead to errors in identification. Never pluck the specimen from the soil or other substrate, as important structures below the level of the leaf litter, turf or soil may be lost or damaged; instead, dig it out carefully. A strong pocket-knife, pen-knife, kitchen

LEFT **Mushroom collecting for food requires good, clean specimens; old or mouldy specimens can lead to upset stomachs.**

knife, fern trowel or similar tool is very useful for collecting specimens. Before leaving a site, always make a note of what the fungus is growing in or on – for example, 'on horse dung', 'in moss', or 'on herbaceous stems'. Include, if possible, the name of the species or genus of plants, especially trees, with which it is associated. A leaf from the tree or other plant, or even a piece of bark, will act as a useful reminder when back at home or in the laboratory, and will hopefully allow the host to be identified later if this was not possible in the field. Such observations are very important, as the more knowledge mycologists can gather the more they are finding relationships that were not previously appreciated. Unfortunately, many early collectors did not realize the significance of such ecological information.

Taking records in the field

Note any unusual features about the fungus, such as the identity of any distinct smell, whether or not the colour of the fruiting body changes where it has been handled or bruised, the presence of hairiness or stickiness, and so on. These can be written in a field note-book, but a pocket tape-recorder is in some ways even better, as the fungus, its substrate and the surrounding area can be quickly described and the recording played back later on when one can retreat to a more favourable location, as one might need to do to escape a tropical downpour. A photograph of the site and the surrounding locality are useful additions to the data.

Transporting your finds

It is important to return to home, camp or laboratory with your collections of material in as perfect condition as possible. For this, an assortment of small plastic or cardboard tubes or tubs and metal tins is essential; waxed paper or even kitchen foil is particularly useful for fleshy agarics. It is not necessary to buy expensive equipment as you can easily recycle readily available food and other containers. One of the benefits of this is that when they become damaged or contain

ABOVE **The old, now disused method of preparing fungal herbarium specimens was to slice them, as in this example. Now, specimens (unless extremely bulky) are dried in a hot air oven at 40°C in their entirety; bigger specimens may be cut in half.**

spores of a previous collection, they can be recycled and easily replaced at little or no expense. The use of glass-tubes is *not* recommended, in case of a fall, which can happen all too easily, breaking the tubes and possibly damaging the specimen – or the collector, which could be a serious matter, especially if one is collecting in a wilderness area miles away from the nearest habitation.

Your selection of collecting materials is best kept in a basket: collapsible types are easier to carry if travelling long distances. In such difficult terrain as is often encountered in deserts, mountains or arctic areas, an upright haversack, equipped with several small chests with assorted drawers that are readily available from hardware stores, is very convenient. However, such backpacks are not recommended when collecting in dense undergrowth such as rainforest, however, as they can easily become caught up in lianas and rattans; a basket is far more manoeuvrable.

Any larger, fleshy specimens should be carefully wrapped in the waxed paper or aluminium foil, and the smaller ones placed in the tubes and tins where they can be accompanied by some of the moss or leaves with which they are associated, to act as a source of moisture. Whereas the larger fungi have sufficient water to keep them turgid after being severed from their mycelial connections, the smaller ones can quickly dry out and shrivel. The woody, tough specimens can be placed in a selection of strong brown paper

Taking a spore print

If their fruiting bodies are placed face down on a piece of white paper or a glass-slide, all the larger fungi can produce a pattern as a result of their spores falling onto the smooth surface. These are called spore-prints and result from the spores being forcibly discharged onto the surface below. In nature these same spores would be ejected into the turbulent atmosphere around the maturing fruit-body to assist in their dispersal.

If the instructions given on the previous pages have been followed, your specimens should be in a pristine condition for examination. It is a good idea to take a spore print of any fungus you want to investigate as soon as possible after you return home. If you have more than one example of a species, you can cut off the entire cap, but if you have only a single specimen, remove a small portion only. Place the surface bearing the spore-producing reproductive tissue (hymenium) – gills, pores, teeth or wrinkles – face-down on a glass-slide or on a piece of white paper. You should then return the specimen to its tin, or cover it loosely with waxed paper or a plastic container, cup or jar. Place a small drop of water on the cap of small specimens to stop them drying out; even the largest specimens need protecting from desiccation, so enclosing them is essential; a glass slide slotted between the cap and gills may suffice with the biggest examples.

Leave the specimen in place for twelve hours and you will have a good spore print. After laying a bracket or resupinate fungus face-down on the paper or slide, you should wrap it in damp newspaper overnight as these types are often rather dry when collected in the field; this procedure will give the fungal tissue sufficient moisture to produce spores. Except for the stinkhorns and bird's-nest fungi, which can be dried immediately, stomach fungi are generally best cut in half in order to see if they are mature; ripe spores are necessary for identification and undeveloped fruit-bodies are of little use.

BELOW **The spores cast by the underside of a mushroom is known as a spore-print, as it gives a trace of the mushroom's gills. Poroid fungi will give a pattern of small spots. This example is of a true mushroom, *Agaricus* sp.**

bags, but these can easily get wet after a shower and disintegrate; a few specially made linen or material bags are a much better choice. Plastic bags are not suitable, as the fungi will disintegrate and form a soggy mass, due to their respiration in an enclosed environment during the journey home. Wrapping specimens in paper towels is also of no use, as it is often infuriatingly difficult to separate fungus from paper.

Recording colour and form

When you have obtained your spore-print, determine the colour accurately by scraping the spores into a small pile and then placing a glass cover-slip on the top; thin deposits can give a very false colour. If you find judging the colour difficult, try painting a swatch of the same colour and comparing the two; this can be a great help. Many different species, although recorded as having simply white, brown, pink, or black spore deposits, actually exhibit further subtleties of hue and these can help to distinguish genera or species.

Once you have described a specimen, you should take a colour photograph of it, or make a colour painting or sketch: water soluble, blendable crayons are excellent for such studies. In some ways, artwork is better than photographs, as it allows you to make a composite representation of a species, based on several specimens. It may be necessary to dry your fungi when it is not possible to identify them immediately, for instance when you are many miles away from a location suitable for processing specimens or with a range of identification books. Many larger fungi contain over 90% of their fresh weight

in water. Fleshy fungi will therefore soon putrefy, while leathery ones may harbour beetle and fly larvae that will feed on the fungus and destroy it from the inside. Thus any specimens not quickly examined can lose even the characters which may enable you to identify them in their dried state.

BELOW **Good colour sketches or photographs are preferable as a fungal record to accompany the dried specimen; extensive fieldnotes are essential as fungi change dramatically once dried.**

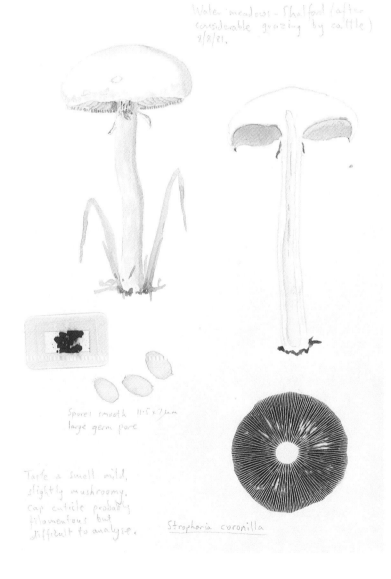

Water meadows - Shalford (after considerable grazing by cattle) 9/8/81.

Spores smooth 11·5 x 7µm
large germ pore

Taste & smell mild, slightly mushroomy. cap cuticle probably filamentous but difficult to analyse.

Stropharia coronilla

Drying fungi

If the luxury of an oven or even a domestic fruit-drier is available then you can dry your specimens at 40°C (104°F), with plenty of air circulation to drive off all the water. A small tray on the top of a radiator or a water boiler can be useful, unless provision can be made to carry small generators to drive an electric drier. If none of these are available, then immerse the specimens in silica gel in a self-seal polythene bag. Try to use 'self indicator' silica gel (unfortunately the blue/pink type is now thought to be a health hazard, but there are other kinds available on the market). A coarse mesh gel is preferable as round granules, by their grinding action, can damage specimens irreparably. If the fungus is large then it can be sectioned, as the field notes taken will indicate the overall features. Leave overnight and replace with new silica gel the next morning, recharging the used gel by heating it in a conventional or microwave oven if this is available. If you are collecting in a remote habitat such as rainforest, you can dry the gel by placing it near the camp fire after breakfast. Repeat until there is no change in the gel's colour. Specimens usually dry overnight but for final storage, place the fungus with a little silica gel in a fresh polythene bag accompanied by your reference notes. You can then rest assured that the material is as fully documented as possible until you are able to transport it home.

Macroscopic examination

Usually, the smaller agarics will give the quickest spore-print. The more experienced collector will recognize species and genera but still needs to examine the spores under the microscope. In such cases, a thick print is not required as sufficient mature spores will be obtained in a hour or so, and a glass-slide is best for obtaining results.

While the spore-print is forming, write full notes on all observable features, such as colour, texture, stickiness, shape and odour. There is no need to become involved with microscopic examination at this stage because it is not necessary, and time is at a premium as changes may occur while the specimen begins to dry out. All changes which take place after collecting need to be recorded. A sketch can often reveal more about a fungus than many words. For future comparison it is always best to have a set format in which all the characters are covered in the same order each time: for example, in the case of an agaric, the sequence could be: cap, stem, flesh

BELOW **Sections of agarics showing the different ways the gills are attached to the stem.**

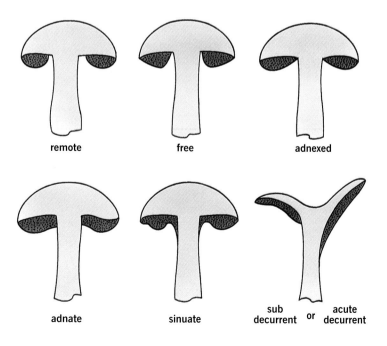

remote   free   adnexed

adnate   sinuate   sub decurrent **or** acute decurrent

LEFT **The bracket fungus,** *Ganoderma collosum*, **broken to show the layer of tubes beneath the cap; St. Lucia, Caribbean.**

(with any associated colour change and odour), and gills, especially the attachment to the stem, which is not visible by peering between the gills and can be seen only by making a longitudinal section through the centre of the cap. The different types of attachment are shown in the diagram opposite.

It is at this stage that any macroscopic tests suggested by the appropriate texts should be conducted.

For detailed examination of macro-characters all that is required is a x 10, x12 or x15 hand lens or even a pair of jeweller's or philatelist's goggles, which have the advantage of leaving the hands free to manipulate the specimen. A dissecting microscope, although useful, is not essential.

Microscopic examination

Techniques of microscopic examination cannot be covered in detail in this book, so the reader should consult the bibliography for appropriate texts dealing with this aspect of identification. However, it is essential to measure the spores, and to note their colour in water, their shape, ornamentation, whether holes and appendages are present, and the thickness of the wall. There are also colour changes that occur when the fungus is treated with specific chemicals that should be noted. It is necessary to examine the structure of a thin slice (known as a 'scalp') of the cap cortex, and determine the shape of the outer cells that make it up, and also to look at the flesh and see if it is composed of one type of filament or a mixture of packets of rounded cells and filaments. Finally, if your specimen is of an agaric, you should make a longitudinal section of the gills, because the arrangement of the internal tissue is diagnostic. As the diagram on this page shows, the internal tissue may be regular in structure and then straight, convergent from the base upwards, divergent from the top of the cap tissue, irregularly arranged, or made up of two distinct zones. There may be sterile cells, called cystidia, often having very distinctive shapes, on either the face of the gills, pores or other spore-bearing structures, or at their margin, on the stem and the cap, or on any combination of these – or, in a club or resupinate fungus, scattered on the free surfaces.

As you gain experience, it will not be necessary to look at all of these features, but when collecting in a new region then even the most knowledgeable mycologist might be very disappointed if the simple requirements outlined above are not followed.

Microscopic work inevitably requires an outlay of money, but this need not be prohibitively expensive. A relatively cheap compound microscope is all that is required to see the major structures, while a modest sum will purchase the rest of the equipment needed: a good supply of glass-slides and cover slips, a pair of fine forceps, some single-edged razor blades, a camel-hair brush for picking up or moving small sections, and a pencil with an eraser at the end that can be used for gently tapping out the tissues and removing unwanted air-bubbles.

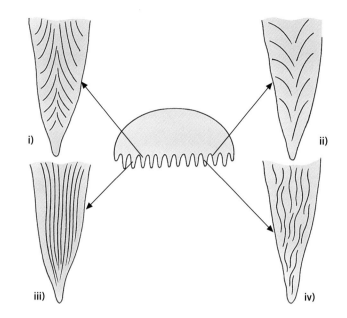

i)

ii)

iii)

iv)

BELOW **Section of a mushroom, showing where the section is taken to inspect the gill-trama. Trama types: i) divergent; ii) convergent; iii) regular; iv) irregular.**

Fungi and humans

In the United Kingdom, in contrast to many other parts of Europe where the eating of larger fungi has a long tradition, fungi have become more popular as food comparatively recently. This is probably because of media coverage and the fact that more British people are travelling abroad, where fungi, other than the cultivated mushroom *Agaricus bisporus*, are regularly eaten. This trend has been seen in other anglicized communities scattered throughout the former colonies, and especially in eastern and southern Europe. The incorporation of local fungi into the diet has soon followed such movements of population. The same has happened in the United States with the immigration there not only of people from Eastern Europe but also those from Japan and Taiwan, and more recently Vietnam. There are, however, only a limited number of species that are popular in Europe, including the chanterelle and penny bun (known in various countries by many different names, including cep, porcini or steinpilze), which are considered delicacies and these can generally sustain localized collecting, both for personal use and commercial exploitation.

Fungi as food

Worldwide, many fungi are eaten, but the choice of species collected for the table in a particular country often depends on religion and culture; indeed a species eaten by one tribal group may not be eaten by another even though geographically the groups may have settled quite close. Sometimes, because of the bizarre shape of a fungus, it may, although perfectly edible, become forbidden food because of a taboo. The fungi used as food range from the enormous variety seen at the height of the collecting season in Thai markets to a highly specific selection in East African countries. In some parts of the world, there are even fungus festivals held at the fruiting time of a particularly sought-after

LEFT **A typical southern European preserved mushroom display showing dried porcini, *Boletus edulis*; Stresa, Italy.**

kind of fungus, such as that for *Termitomyces* species in north-west Thailand.

Some of the mushrooms eaten in Southeast Asia are now grown commercially and appear on a regular basis in Western shops and supermarkets. None approach the massive production of the white cultivated mushroom, which worldwide is estimated to be worth millions of pounds.

Recently, a closely related brown cultivated mushroom and the Portobello mushroom (another agaric) have come onto the market on a big scale. The appetite for other species supports an ever-increasing trade. First to appear was the oyster mushroom, of which three species are now readily available, the white oyster, pink oyster and yellow oyster; these were soon followed by the black forest mushroom or shiitake, *Lentinula edodes*.

More recent additions are the velvety shank (or enoki), *Flammulina velutipes* and lion's mane fungus (or bear's head fungus), *Hericium erinaceum*. In other parts of the world the Jew's ear fungus is grown, and in Europe attempts have been made to introduce the king stropharia, *Stropharia rugosoannulata*, blewits and wood blewits (*Lepista nuda)*, the first not all that successfully.

All these commercially produced species are grown in artificial culture but the most popular fungi (apart from the commonly bought cultivated mushroom, *Agaricus bisporus*) in the western market are chanterelle, penny bun and, most prized of all by far, truffles. The world consumption of the first two is estimated at 200,000 tonnes per year. These three fungi are all collected from the wild as they have not been successfully grown under artificial conditions; they all rely for growth on an ectomycorrhizal association with a tree. The crop is either sold fresh, when it must be quickly transported to the point of sale before deterioration, or is sliced and then dried. The former method is the basis of a million-pound operation, with the fungi coming from many parts of the world; the drying is a relatively low-key operation, as fungi so preserved are used for incorporation into dried soup or for sale as additives to be reconstituted in water before being added to stews, risottos and other meals.

Packets of dried wild mushrooms composed of a whole selection of native European species, including the slippery jack, are now popular.

It is very difficult to produce an appetizing product after drying cultivated mushrooms, as they darken unattractively when so treated, thus making the other, wild, species more popular. Cultivated mushrooms are instead preserved by pickling, and less frequently canning. At certain times of the year in Britain winter chanterelle, champignons mouton and horn-of-plenty can be purchased. In other parts of the world other wild species are frequently collected and sold; the collection of matsutake for sale in the north-western United States actually provoked a murder over control of the best collecting spots.

In Asian and African markets there is a product termed vegetable cheese which is produced by allowing some easily available source of naturally occurring carbohydrates (such as cassava, also known as manioc) to become colonized by a filamentous fungi, the whole product then being eaten. In China, soya is used and the product is called 'sufu', and in Japan 'tofu'. In western society this is paralleled by Quorn, a fungal protein (or mycoprotein) food that is increasingly eaten by vegetarians and others, too.

Truffles

Fungal structures, both above and below the ground, are a source of food for a variety of animals. Fruiting bodies are often devoured by the maggots of beetles and of flies (which may in turn be eaten by invertebrate carnivores), as well as by mammals, especially mice, deer, squirrels (including chipmunks in North America), and by bandicoots in Australia. Indeed, some fungi, notably the truffles (which are ascomycetes) and the superficially similar but unrelated false truffles (which are basidiomycetes), have evolved intimate relationships with certain mammals. The mammals regularly seek out and dig for these subterranean fungi, attracted by their special aromas. Indeed, the best way of successfully finding truffles and other subterranean fungi in the field is to search where small mammals have been digging; a good mycologist is always a good naturalist!

RIGHT **Various mushroom products including both dried and canned varieties as well as labels from pickled mushrooms and paté containers.**

The fruit-bodies of these fungi can grow as big as a large potato, carrying out their entire life-cycle hidden from view; their spores are eaten with the fleshy parts of the fruit-body and dispersed far and wide after having passed through the gut of the mammal and ejected unchanged in its dung. In some parts of the world subterranean fungi form a major part of the diet of small mammals, while of course many humans regard them as a delicacy. A special breed of dogs (Dorset hound) has been trained to hunt out truffles although more commonly domestic pigs are used to search for this purpose. The pig used is always a sow, as the smell of some species of truffle resembles that of the male pig.

Because of their shape and smell, truffles have been accorded the status of a valued aphrodisiac, and therefore large, undamaged specimens can demand a very high price. At the time of writing, this is regularly in the order of £100 or more per fruit-body, which equates to £300–£500 per kg (£136–£227 per lb) for the winter truffle and £1500 per kg (£682 per lb) or more for the white truffle – in 2001 even as much as £2000 per kg (£909 per lb); the record is £22,000 for a 1 kg (2.2 lb) white truffle sold at auction in California, USA in November 2002. There is quite a substantial trade in truffles throughout Mediterranean Europe, but more recently exports are coming from China. Although truffles are ectomycorrhizal and therefore dependent on a tree association, they have been successfully grown under introduced pines in New Zealand.

Destructive species

In addition to the losses from fungal diseases of crops, expensive and often devastating damage is also caused by macrofungi.

Dry rot

The scourge of many houses in Europe and North America is the dry rot fungus, a relative of the bracket fungi. Its scientific name *Serpula lacrimans* refers to two distinctive features of the fungus: the generic name *Serpula*, from the Latin word meaning 'a little serpent', relates to the sinuous shape of the spore-producing surface, while the specific name *lacrimans*, from the Latin *lacrima*, 'a tear', refers to its ability to produce tear-like drops of water when it is actively growing. Its common name, however, can be misleading, because it needs some moisture to start growing. However, when it is well established, it produces water as it decomposes the wood, which becomes brittle and dry; it is probably this that accounts for the name.

BELOW **Summer truffle,** *Tuber aestivum***, an avidly sought truffle with good texture and flavour, and above all odour.**

RIGHT **The dry-rot fungus,** *Serpula lacrimans*, **fruiting on the door-frame and in the roof cavity of a house.**

Dry rot is able to spread so effectively because the mycelium does not remain in the wood first attacked, but forms remarkably tough strands that snake across all sorts of structures that provide no nutrient, until it reaches a new area of wood on which to feed. These strands can pass through tiny gaps and pores in brick, stone, mortar and plaster, beneath floors and anything else along which it can grow, including telephone cables and the like. When the mycelium can grow no further in one direction, as when it reaches the outer surface of a wall, it may form a fruiting body, which may be bracket-shaped and can cover several square metres. At first, a fruiting body is thin and fluffy, but as it grows it becomes leathery and wrinkled, varying in colour from dull yellowish-grey to rusty brown, with a wavy, raised or swollen, whitish margin. Each fruiting body produces many millions of spores, that can give rise to new infections.

As long as the snaking mycelium can find new wood to attack when its original supply is exhausted, the fungus can continue to spread, at first insiduously and often unnoticed. Another reason why dry rot is so destructive is because it does not just spread out laterally; it is cubic in nature, the fungus attacking the wood in three dimensions. If it is not controlled, dry rot can cause appalling destruction, pervading an entire house. Walls bulge and split, the timber is discoloured and softened and, as the wood dries, it is eventually reduced to a crumbly powder, so that floors and staircases give way underfoot.

Widespread for centuries in the West, the dry rot fungus was almost certainly introduced from the Himalayan region. Historical records certainly show that it has been a menace for a long time. For instance, it is known to have damaged the *Queen Charlotte*, a 110-gun battleship launched in 1810, which rotted so quickly it was necessary to rebuild her before she could be commissioned for sea. In 1859 after extensive repairs she was renamed *Excellent*.

Wet rot and others

Less damaging is wet rot, caused by the wet rot or cellar fungus, *Coniophora puteana*. The wet rot fungus is often called by its synonym *C. cerebella*; it is classified in the same family (Coniophoraceae) as the dry rot fungus, but differs in that it attacks only wood with a high moisture content, accounting for the common name. Wet rot often causes damage to wet cellars, door and window jambs, garden sheds, out-buildings and other sites when the wood becomes soaked, producing a tell-tale dark stained, longitudinal rot.

Different fungi can be found in houses and other buildings, especially in damp rooms and cellars. A good example is one of the elf cups, *Peziza cerea*, whose fruiting bodies commonly appear on concrete pointing between stonework, wet sand-bags and even in the corners of outbuildings and bedroom ceilings. Even more alarming is the appearance of ink caps and even oyster mushrooms in damp areas of rooms or on wooden boarding at the back of kitchen-sinks.

In greenhouses, where the environment is warmer, damper and more constant there is a whole range of unwelcome but ubiquitous

bracket fungi that can rot wooden staging and seed-boxes. Other large fungi may grow in compost and some quite often fruit in pots of house plants purchased from nurseries or retail shops.

In various Western countries before the Second World War other large fungi caused many thousands of pounds' worth of damage and endangered people's safety, as with the wholesale rotting of wooden sleepers on railway tracks by *Lentinus lepideus* and of pit props supporting the roofs in coal mines by *Antrodia vaillantii*.

Larger fungi rarely cause spoilage of food or animal feed, although the split-gill fungus damages hundreds of tonnes of silage annually in the UK, rendering it unfit for feeding to livestock. This is a relatively recent

LEFT **The wet-rot fungus, *Coniophora puteana*, is found in both natural habitats such as woodlands and in buildings on wet timbers.**

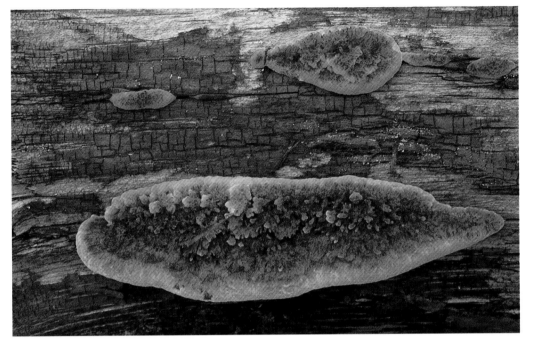

LEFT **The bracket fungus,** *Gloeophyllum sepiarium*, **grows on coniferous wood both in the house and garden; Argyll, Scotland.**

phenonemon, which has been linked to a change in farming practice, in which the grass or other herbaceous material is nowadays fermented within black polythene bags instead of in a silage pit or silo.

By contrast to these tales of destruction, in North America it has recently become popular to use what is called 'pecky wood' in houses; this refers to lengths of timber used as decorative features, in which the wood-rotting fungus has produced differently coloured patches, and sometimes even holes, that are considered attractive.

Problems in the tropics

The hot, damp environment of a greenhouse is similar in many ways to that found in the tropics, where there is a whole range of bracket fungi and their agaricoid relatives that rot man-made structures. Wood is used far more extensively in place of metal and other fungus-resistant materials for buildings in developing countries compared with those in the industrial West, and the extensive damage that is still regularly caused in the tropics is a problem, especially as chemical protection is expensive.

The association of insects and fungi in the tropics is generally much more apparent than in temperate areas. Termites are rapidly attracted to timber as soon as any colonizing fungus gives off aromatic or similar volatile compounds, as by-products of wood decay. These chemicals act over long distances like the olfactory equivalent of a smoke-signal to the termites, which soon appear and multiply many times the damage caused by the fungus as they devour the wood with their powerful jaws. Sterile wood in sterile soil fails to attract the termites, so treatment of the site with poisons is paramount to stop fungal colonization. Phenolic compounds were originally used, but there are also many substitutes available today; vast amounts of money are spent annually in treatment and on research aimed at finding better ways of protection.

Other relations with animals

Domesticated animals such as sheep, horses and cattle also eat mushrooms, again dispersing the spores in the dung. Some animal behaviourists have proposed that other animals, such as dogs, may apparently even seek out fungi for purposes of self-medication. Unfortunately, they sometimes misidentify the fruit-body and are poisoned.

There are other groups of fungi, such as the stinkhorns, that are dependent on insects for the dispersal of their spores. In the soil, fungal spores and the mycelia of both larger fungi and microfungi provide food for microscopic soil organisms, such as amoebae, flagellates and other protistans. Some amoebae eat the bacteria that adhere to the hyphal walls, where the bacteria in turn are feeding on nutrients leaking from the fungal hyphae. Minuscule nematodes (eelworms) and little springtails also feed on fungi.

Poisonous species

Although the production of edible larger fungi is a multi-million-pound business, far more publicity is given to poisonings of humans by wild mushrooms. Despite this bias, it is always vital to take the greatest care when collecting this source of food, as some widespread fungi may be very poisonous.

Care is always essential when eating fungi, because many people who have been unaffected by eating certain types of mushroom for years can suddenly become allergic to them. Try only a small portion if it is the first time you have eaten the fungus. A sensible rule of thumb is never to eat fungi of whose precise identity you have the slightest doubt, and always to consult an expert in the field for a correct identification.

This is particularly applicable to less experienced collectors or newcomers who are unfamiliar with a country. Even with a good knowledge of the edible species in their homeland, people who indiscriminately eat similar looking fungi in their newly adopted home can put themselves at risk. Eating the wrongly identified fungus may result at best in a badly upset stomach or at worst damage necessitating a kidney transplant or even a fatality. It is not true that animals know better, so seeing a fruiting body nibbled by a slug or a mouse is no proof of edibility!

There are generally more poisonings in the areas of the world where more species are regularly eaten than in 'mycophobic' countries such as the United Kingdom, where one poisoning or even the sighting of a potentially dangerous fungus makes headline news. In Britain there are few serious poisonings annually and there have been no recent fatalities. The most common fungal poisonings are those of small children in the early summer, when they are crawling and playing on lawns in gardens and parkland; the fungus often involved is the haymaker,

LEFT The death cap, *Amanita phalloides*, is found throughout the British Isles especially old oak forests. It is the most poisonous British agaric.

Panaeolina foenisecii. Although this species is not particularly toxic for adults, as always, poisoning is a balance between body size and the strength of the toxin involved.

In the United States a hot-line and register of fungal poisonings is kept as there are more poisonings per head of population. Unfortunately we possess little data for most developing countries. Generally native peoples, at least in rural areas, are good at fungal identification and train their children in recognition from an early age.

Hallucinogenic mushrooms

There is a small group of larger fungi that induce dramatic changes in perception and mood and hallucinations; the fungi involved are usually in the genus *Psilocybe* and the chemical compound causing the effects is psilocybin, which is closely related to lysergic acid, a building block of the famous drug

LEFT The liberty cap, *Psilocybe semilanceata*, a very common 'magic mushroom' found in gardens, parks, hill-pastures and playing fields throughout the northern hemisphere.

73

LSD. The commonest species is the so-called 'magic mushroom' collected wild in grassy places, often near to habitation, in many temperate parts of the world. 'Mexican gold-tops' occur wild in subtropical and tropical countries; this larger species *Psilocybe cubensis* is being successfully but illegally grown in Western countries.

Hallucinogenic fungi used for recreational purposes have become exploited only relatively recently in Western society but they can be easily misidentified. Cases are known of ingestion leading to hospitalization; over-consumption can lead to poisoning and in some people the recurrence of a series of unpleasant symptoms. In the Far East other species, such as 'blue meanies' *Copelandia cyanescans*, related to the haymaker, have

been used as recreational drugs. These same fungi have been the basis of religious ceremonies in many parts of the world in what can only be unconnected cultures. In Central America, for instance, so important have hallucinogenic mushrooms been in the past that stone effigies have been made of them with their accompanying god. In the far northern parts of Europe and Asia similar fungi have been involved in shamanism and in ceremonies for tribal leaders; there have even been suggestions that some modern religions have their origins in mushroom worship.

Unusual uses

A rather more recent development is the utilization of the very strong hyphae of some poroid fungi for making paper. The process is similar to that using wood pulp to make handmade paper on a small scale. Although this particular application has only a short history, eastern Europeans have over many generations soaked and beaten out strips of bracket fungi to mould into household utensils and clothing accessories. Today, hats, belts and bags made from the tinder fungus are an important tourist attraction in some small towns in Hungary.

Early humans were aware that these same strong hyphae could be used to sharpen the edges of weapons and tools as indicated by the pieces of fruiting body from at least two different bracket fungi, found in the possession of an Ice Age man found in the Alps in 1991. Even today, cut-throat razors are still sharpened on a 'leather' made from poroid fungi, and these poroids make

RIGHT **Guatemalan mushroom stone in a private US collection.**

RIGHT **The birch bracket or razor strop fungus,** *Piptoporus betulinus*, **is very common in European birch woods and has been used as its common name suggests to sharpen knives.**

excellent material on which to dry fishing flies. It is not long ago that strips of brackets were placed in museum cases to stop exhibits from becoming mouldy.

The use of these fungi as totems has been identified in the culture of North American Plains Indians, where pieces of a sweet-smelling bracket have been found sewn to garments. Similar larger fungi have also been used as fragrant decorations in the past, and a range of bracket fungi are now sold in markets and florists for use in floral and other decorations.

In some rainforest communities the tough bootlace-like aggregations of hyphae which form aerial mycelia have been used as a primitive thread to hold utensils and clothes together. In the same communities pieces of branch or stick found on the forest floor and observed previously to be impregnated with a bioluminescent fungus are used in hunting. The pieces of wood are placed across animal

tracks and when the light given off by the fungus is seen to shift then the hunter throws his spear at the movement of light which indicates the presence of an animal. This is much more sensitive than the hunter waiting until he hears where the prey might be.

In fact, there is a wide range of fungi, especially tropical species, that are known to emit an eerie green or blue light; mycologists have little knowledge of why they do this. On a warm summer night, the honey fungus, a common species found worldwide from the

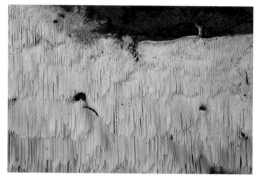

LEFT **The bracket fungus, *Antrodia vaillantii*, although found in natural communities also occurs in coal-mines, domestic premises, and on woody artefacts where it may grow on vertical surfaces.**

LEFT **Bootlace-like structures of various agarics similar to those of the honey fungus, *Armillaria gallica*, have been used by many cultures to join clothing and parts of utensils together.**

LEFT The tinder fungus, *Fomes fomentarius*, common on birches in Highland Scotland but occasionally on other hosts elsewhere has been used in Central Europe to make hats, aprons and widely elsewhere as a source of tinder.

temperate regions to the tropics, gives off a distinct phosphorescence if the bark of the infected tree is stripped off to expose the mycelium.

The versatile tinder fungus

The male agarick, or tinder fungus, is also a bracket; in the past it was made into a wool that was mixed with saltpetre to form an important ingredient of the musketeer's tinder box. This was ignited by the spark produced as the gun's hammer struck the flint, in turn lighting the wick to set off the main charge for firing the shot. This fungus is still used by some north European peoples such as the Lapps, to put on their fires at night to glow and smoulder, keeping wolves and bears at bay. It saves them using up their limited supplies of wood (needed for cooking and heating) for this purpose. The fungus also has the advantage that it can be retrieved from the fire and used again.

Tinder fungus has been recovered from many archaeological sites in Europe, as have puffballs. These apparently were also used as smouldering wicks to carry fire, just as some tribal people in the Himalayas do today. Another use for smouldering puffballs in some societies is to cauterize wounds, while the smouldering fruit-bodies have been used in parts of Africa to smoke out and calm wild bees in order to ease the collection of honey without being stung. Cramp-balls have been used for the same purposes.

Fungi as dyestuffs

Using larger fungi for dyeing is very popular today in many countries with a Western culture, but it is not widespread outside these communities nor was it so in ancient times; the use of larger fungi for this purpose therefore does not parallel the long history of lichenized fungi being used in dyeing.

Over seventy different colours can be produced from larger fungi by changing the mordant (fixative substance) in the wool, the pH, the quantity of the fungus used and time of immersion in the extracting bath. Wool, silk, hair and paper can all be dyed in this way, using various species of fungi from a range of genera.

Whereas today many species of agarics are used, the few traditionally used fungi in the

ABOVE **Yarn dyed with various fungal dyes and two hats knitted from such wool.**

RIGHT **The devil's foot or devil's club fungus, or dyeball, *Pisolithus arhizus*, has been used by medieval Europeans to produce a rich canary pigment.**

past were all various bracket and stomach fungi, that have a more persistent fruit-body. Plains Indians in North America used the woody forest bracket that became known as the Indian paint fungus, *Echinodontium tinctorium,* which they obtained by trading with other native American tribes; from this they produced a rich orange-red pigment that stains the skin. This may account for the name 'Red Indian' given to these peoples by the white colonists. Europeans were introduced to the rich yellow colours of silk dyed with the devil's-foot or devil's-club fungus obtained through trade based in Genoa. The raw material probably originated from Madeira or the Canary Islands, possibly giving the name to the colour canary yellow. Aboriginal Australians used the paint brush fungus to decorate their bodies; the spores gave the effect of a dye when mixed with saliva.

Fungi as medicines

Long traditions of the use of fungi in medicine occur in all cultures of the world, including the use of very dilute solutions of a poisonous agaric *Amanita phalloides,* the death cap, to alleviate illness as practised by religious orders in Europe, in what could be seen as a forerunner of homeopathy.

Both bracket and stomach fungi have been used in medicine for many thousands of years, by communities ranging from Stone Age tribes and later Roman colonists in Europe to Himalayan peoples today. The puffballs used have spores that are about the same size as blood cells and if applied to a wound encourage rapid clotting; others are known to have antibacterial activity.

Bitter-tasting substances occur in some bracket fungi and are probably the compounds involved in enabling wounds to be staunched during surgery; one such bracket used in early Europe was known as the surgeon's agarick.

The female agarick, *Laricifomes officinalis,* is a poroid fungus growing on larch trees that was a source of trade between Britain and Europe in mediaeval times. It was a panacea used by generations of doctors from Greek to Mediaeval times to cure a great range of ills; similar fungi have been identified in other parts of the world. In eastern Asia, especially China, there is a vast pharmacopoeia of larger fungi that may be prescribed in medicine.

The most widely used bracket in traditional Chinese medicine is a species of *Ganoderma* known to the Chinese by the name Ling Chi. It is now grown commercially for Chinese medicinal stores and can be purchased shredded or sliced, as a sort of tobacco for smoking, in the form of a cordial or tonic, and as sweets. The cloud mushroom *Trametes versicolor,* also known as *Coriolus* and *Polystictus versicolor,* appears in oriental pharmacopoeias and has recently been made available in tablet form in Britain.

Microfungi also have a long history of use in medicine. The tradition of placing mouldy poultices on wounds unwittingly made use of the production by the colonizing fungus of chemicals that have antibiotic properties. Nowadays, a whole series of these compounds is available to us, although when food is in storage these same fungi may also be major spoilers of food. Penicillin was famously discovered by Alexander Fleming in

1928 after accidental contamination of an experimental dish of disease-causing bacteria by a *Penicillium* mold, which killed the bacteria, and has saved many thousands of lives since. The top three best-selling pharmaceuticals are derived from microfungi, while similar biologically active compounds make some of the larger fungi important in eastern medicine.

Myth and Legend

The fact that larger fungi, often bizarre in shape, suddenly appear, often overnight, in damp, dark places has created a mystique around them which persists in many cultures worldwide. They are the subjects of many woodcuts, totems and other works of art, and have also been used in religious ceremonies from ancient times.

In literature and storytelling, especially in medieval Europe, fungi were linked with toads and reptiles, and subsequently with bats and nightjars, which both fly at night as witches and wizards are said to do, and were

therefore thought to be connected with black magic. Such unexplained phenomena as the rings of some species such as the fairy ring champignon *Marasmius oreades* that appear overnight on lawns and other grassy areas, were said to be the result of fairies dancing in a circle while humans slept, earning them the charming name 'fairy rings'.

ABOVE **Pattern produced by the activity of the fairy ring champignon,** *Marasmius oreades*, **growing out in all directions from a central point to form a ring, and within which the agaric fruits under favourable conditions.**

Conservation

The conservation of fungi is not as straightforward as that of some of the vascular plants or vertebrate animals. Effective conservation depends on a high degree of knowledge of the organisms found in a particular site, and this is rarely available for the fungi.

As the major part of any fungal system is below the ground and goes unobserved, the presence of a species of fungus in a particular place can be readily established only by recording identifiable above-ground structures. In a particular year, larger fungi may fail to fruit, and this may continue for several years on end. As a result, records of occurrence are often patchy.

Identification of larger fungi is also often fraught with difficulties as the majority of species require careful, often microscopic, analysis. However, the difficulties of identification, especially of some critical species, is gradually being resolved by the appearance of more reliable field guides, including some that cover hitherto little-documented species found in tropical regions.

These two factors are the main problems hampering progress in gathering and analysing field data. This has led many people, even including biologists working in disciplines other than mycology, to consider the fungi as nothing more than 'background noise', and probably having little real effect on the total ecology of a site. To the initiated this has long been patently untrue, but now the enormous role played by fungi in all ecosystems is gradually being more generally appreciated.

Current strategies

At present, conservation strategies for larger fungi are concepts that apply only in Europe and, to a lesser extent, in North America. The few examples in developing countries are generally instigated by western mycologists backed by western finance. Fungal conservation is an expensive activity and at the moment has not been tied to other benefits that would make it more palatable to influential organizations, although as we know from alternative medical treatments they can produce some potentially useful drugs to add to those, such as penicillin, already used for many years in conventional medicine. It is fair to say that, apart from those at a very few sites, the precise range of species of fungi present in any one area and their relative abundance or scarcity is usually unknown and therefore conservation is meaningless.

Habitat loss

Habitat loss seems to be the main factor involved in the disappearance of fungi or threatening the majority of our less common fungi. It is of vital importance to address this problem when trying to protect what we consider to be the rare species of larger fungi. Human activity in many parts of the world has modified the environment in a whole range of diverse ways, many of which date right back to the time when early humans began to hunt systematically, and later to fell trees and clear land for agriculture. Such altered habitats now have their own characteristic fungi that are worth conserving in their own right. Unfortunately, there is no experimental work that measures the sensitivity of fungi to habitat change in a particular site, and at present mycologists can proceed only via sensible hypotheses. The reason for this lack of basic information is that, apart from a few exceptions, it has been almost impossible to identify fungi in their vegetative stages by using the traditional methods described in earlier sections of this book.

This means that a fungus can be recorded as present only if the fruiting body appears, but if it does not fruit this is not evidence that it is not there. Some limited information can be obtained to overcome this problem by monitoring a single site over a period of seven to ten years, but this is very labour intensive. New technology is now available, including techniques of molecular analysis, that may make such studies easier to carry out in the future, but at present these are expensive. This new approach will allow researchers to identify the fungus in its vegetative state in the soil by comparing its molecular structure with that of known fruiting bodies that have been collected in earlier seasons.

Preserving ancient woodlands, or in developed areas, even small stands or single specimens of veteran trees, is important in ensuring the continued survival of a wide diversity of both larger and microfungi. Veteran trees within tropical rainforests are exceedingly important, especially when (as is often the case) the forests grow on poor soils. Larger fungi frequently decay the heartwood of such trees over long periods of time and, without affecting the health of the tree to any great extent, this plays a vital role in creating cavities that are used as breeding and roosting sites for a whole host of animals, from toucans, hornbills and other hole-nesting birds to small mammals, especially bats and flying squirrels. The waste products from such creatures add to the overall nutrient input into the soil in the vicinity of the ancient trees, often encouraging the scavenging mycelia of ectomycorrhizal fungi to capture nutrients.

In grassland and savanna habitats, the destruction of termite mounds destroys the delicate balance of nutrient flow, as the droppings from the termites no longer act as a focus for the colonization by ectomycorrhizal roots, which in turn benefited adjacent host trees. When ancient grasslands are ploughed and reseeded, a whole range of fungi – and many other associated organisms, including wildflowers – are lost. Changes in land management worldwide is the single biggest factor in a decrease in the diversity of fungi.

Pollution

The increased use of high-nitrogen fertilizers to encourage plant growth for human consumption or for feeding livestock has led to the modification of many grasslands in parts of Europe, with an associated decline in the fruiting of many fungi.

Pollutants drifting into watercourses feeding woodland can have disastrous effects on fungal diversity: research has demonstrated that fertilizers or atmospheric pollutants can reduce the normally great range of fungi that form sheathing mycorrhizas on the roots of trees, replacing them by just two or three common, widespread and apparently successful species. Also, some larger fungi can accumulate high levels of heavy metals in their mycelia. This has the effect of concentrating toxic material in the upper soil layers, where it may remain and damage roots of trees and other plants. However, it may be possible to enlist the aid of some of these fungi in the biological 'cleaning' (bioremediation) of water from some pollutants.

It is advisable to avoid eating edible fungi that have been collected along busy highways, on industrial sites or in similar localities, as there can be considerable accumulation of heavy metals, including radioactive nucleotides (unstable atomic nuclei), within the fruiting body, posing a potentially serious health risk.

It is not widely appreciated that fungi themselves can produce compounds that we consider pollutants. About 160,000 tonnes (159,000 tons) per annum of chlorinated greenhouse gases are produced by a single family of wood-rotting fungi alone. Whereas originally these compounds would have been part of natural cycles, with changes in land use more wood is made available for the fungi to rot, in turn leading to the release of more

RIGHT Range of caesium-137 in various fungi from Central Scotland measured in units of becquerels per kilo of mushroom. Caesium-137 is potentially carcinogenic and mostly originated from the Chernobyl disaster. As shown, some fungi accumulate high levels. (BG = background level)

Common name	Species	Life strategy	Cs^{137} content (Bq per kg)
amethyst deceiver	*Laccaria amethystea*	ectomycorrhizal	115.5
beech emetic	*Russula mairei*	ectomycorrhizal	251.6-1011.3
blackening russule	*Russula nigricans*	ectomycorrhizal	107.4-395
birch polypore	*Piptoporus betulinus*	on wood	103.6
blusher	*Amanita rubescens*	ectomycorrhizal	45.5
brown roll-rim	*Paxillus involutus*	ectomycorrhizal	60.5
butter mushroom	*Hygrocybe pratensis*	on soil and litter	113.5
chanterelle	*Cantharellus cibarius*	ectomycorrhizal	133
common earthball	*Scleroderma citrinum*	ectomycorrhizal	497.8
golden sock mushroom	*Phaeolepiota aurea*	on soil and litter	BG
green russule	*Russula aeruginea*	ectomycorrhizal	251.6-1011.3
honey fungus	*Armillaria gallica*	on wood	28.3
pelargonium-smelling russule	*Russula fellea*	ectomycorrhizal	97.7
penny bun	*Boletus edulis*	ectomycorrhizal	68.4
poison pie	*Hebeloma crustuliniforme*	ectomycorrhizal	BG
shaggy pholiote	*Pholiota squarrosa*	on wood	BG-138.4
wood woolly foot	*Collybia peronata*	on soil and litter	94.7

of these polluting gases into the atmosphere. As such a profound effect can occur with just one family of larger fungi, it should be of concern that we are ignorant of the roles members of many of the other families play in ecosystems.

Red data lists

In Europe particular attention is being paid to vulnerable and endangered fungi, and National Red Data lists for some countries have been or are being prepared in response to the 'Earth Summit' (United Nations Conference on Environment and Development) held in 1992 in Rio de Janeiro. In addition, the World Conservation Union, also known as the IUCN (International Union for Conservation of Nature and Natural Resources) is trying to accommodate fungi within its major management initiatives. The UK Biological Action Plan has been formulated to include a list of species of larger fungi that must be mapped and monitored to define long-term solutions for their conservation. Of the fungi included over a third are ectomycorrhizal species. The stipitate hedgehog fungi, an ectomycorrhizal group, and the saprotrophic waxcaps are two groups which have been chosen for this special attention.

In the course of undertaking these specific action plans in the UK, far more sites have been surveyed for fungi than ever before, including many areas that had never been visited by mycologists, despite the long history of research into fungi in Britain. Except in Denmark, there is nothing similar in place in other countries to map the distribution from field data to ensure the

recording of the fruit-body numbers and phenology, although there are often other specific activities. In Finland special attention is being paid to monitoring primeval woodland while in various other countries there is a focus on determining indicator species to help mycologists assess the relative importance of a site for conservation. An encouraging sign is that, in general, people all over Europe – even in Britain, well-known for its tradition of suspicion towards fungi – are becoming more aware of the vital roles played

ABOVE **The parrot waxcap,** *Hygrocybc psittacina*, **a species which in many European countries is included in conservation studies.**

by this group of organisms. In Australia the national Fungimap system is helping conservationists in this vast island continent to collect some baseline information on the distribution of selected species, some of which may form an important part of the diet of some of Australia's threatened mammals. Rather more ambitious national mapping schemes have been devised by several European countries, the Netherlands leading the world in this field. Surveys of smaller well-defined areas are now being conducted by a whole range of small groups of enthusiastic mycologists all over Europe.

In North America the problems have not been widely addressed, although maintaining the population of spotted owls threatened by the felling of old growth conifer forest has brought attention to the importance of fungi in the forest ecosystem. The owls feed on small mammals, which in turn feed on and distribute larger fungi, many of which are ectomycorrhizal species. The research and conservation initiatives thus demand input from mycologists as well as ornithologists and others, and show how vital it is to maintain the integrity of all members of the ecosystem to sustain the health of the forest.

Many of the larger fungi are specific to a particular host or its remains after it has died. Such associations can be remarkably localized; for instance, among the saprotrophs, some are associated only with leaves, some with wood and so on. Even more precisely, those on the leaves, for example, may fruit only on the veins, or on the blades, or on the stalks (petioles), while those restricted to wood usually fruit only on decaying wood fragments from a particular species of plant, or on branches of just a small range of diameters. The loss of a host is therefore paramount, and any conservation activity must take these factors into account.

In all parts of the world fungi are eaten by wild animals, and this is an important stimulus for the fungus to produce further fruiting bodies as replacements. Conservation activities must always take into account the close relationship between larger fungi and the flora and the fauna of an area.

The need for public education

Larger fungi do not receive the attention they deserve in conservation strategies. A major reason for this is that they produce fruiting bodies only at certain times of the year and often fail to fruit at all in certain years – even though mycologists may consider the conditions favourable for fruiting.

Education of the general public – and biologists working in disciplines other than mycology – is paramount. Rare or unusual finds of larger fungi or visits of fungal specialists to a particular area or another country often catch media attention, but this is invariably short-lived. However, in the UK the Biological Action Plan programmes have made it possible to raise the profile of fungi with both the biologist and the general public. Also in the UK, where vast areas of land are in private hands, a manual entitled *Managing your land with fungi in mind* has recently been published (see Further Information, p. 96).

Education relating to the need for conserving fungi and the importance of what fungi do in nature, especially their interdependence with trees, has an added importance in the tropics and the poorer parts of Eurasia, where larger fungi are also an integral part of the diet. Sustained harvesting must be assessed and monitored, and any exploitation recognized before irreversible damage has been done.

Picking mushrooms does not generally create a threat to the species as replacement fruit-bodies are already in place under the surface and ready to grow. The removal of the dominant fruit-body by picking will stimulate their growth, provided the weather conditions remain favourable. The United States was the first country to conduct scientific assessments of the effects of picking edible fungi.

Codes of conduct

Although it is true that many larger fungi are eaten by wild animals, this is a fairly random activity, unlike commercial picking. In order to try to control the detrimental effects of picking, in particular the side-effects of soil compaction, some countries have introduced codes of practice; different countries have tackled this in different ways. Indiscriminate trampling is perhaps the main damaging factor, as it destroys the underlying mycelium in the soil by changing the water movement and by collapsing the air spaces. In areas where truffles and other subterranean fungi are dug from the ground, replacing the turf or moss removed from the hole is essential to reduce desiccation and the disturbance of the ground layer; this also causes less disruption to the invertebrate fauna. It is also essential to replace overturned logs or fallen boughs when looking for wood-rotters, and to preserve the

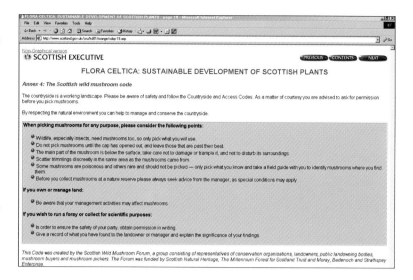

environment in general as close to its original state as possible.

Several northern European countries have published codes of conduct for collecting, and in Sweden the presence of rare and red-listed fungi in sites is accepted as a good argument for the protection and the scheduling of forest and nature reserves; already, five species have been protected in that country by national law.

Nature reserves are the most common, effective and most flexible way of protecting fungi and they also have the advantage of protecting other organisms at the same time. Whereas some nations focus on the protection of rare fungi, others, such as the Netherlands and Switzerland, base their legislation on either the number of fruiting bodies or the weight of specimens collected – a mechanism that ensures that common species are covered as well as uncommon ones. In the United States restrictions on collecting fungi are imposed in areas of scientific importance, national parks and reserves and this may

ABOVE **Extract from the Scottish Code of Conduct for picking mushrooms.**

Specific aspects of conservation

The conservation targets outlined below require emphasis in any education programmes. They are applicable worldwide, although different habitats have different specific needs, as follows.

• In woodland and forest: ensure there is a continuity of tree ages so that saplings and old and ancient trees are present; leave standing and fallen wood and discourage the collection of leaf-litter and wood for burning; maintain a mosaic of understorey types; where grazing is a tradition, try to maintain a range of different grazing regimes, to produce a corresponding range of heights of vegetation beneath the canopy, from long growth to closely cropped areas; maintain old tracks as far as possible, but when creating new trails always bear fungi in mind; prevent or control woodland or forest fires; discourage charcoal production in sensitive areas, such as in *Miombo* woodland in Central Africa, where trees are felled to produce charcoal and create grazing areas for cattle.

• In grassland: try to maintain old grazing regimes and keep the sward short; if the sward is cut, then remove hay or cuttings; do not plough; do not re-seed or supplement the sward; maintain the drainage pattern wherever possible, but when this must be changed, then always consider how changes in the movement of ground water are likely to affect the fungi.

• In savannah and parkland: restrict shade from trees and control shrub development.

• In moorlands, tundra and mountains: prohibit changes in draining; prohibit improvement with fertilizers; do not plough or re-seed.

In all these ecosystems:

• Try to avoid soil compaction by human trampling and overstocking, or from movement of heavy machinery.

• Prevent forest and woodland, grassland and moorland fires unless controlled burning is part of the historic land management practice, and then carry this out at the traditional time of the year.

• Do not use artificial fertilizers or lime to improve plant cover.

cover common, widespread species; permits to collect non-edible kinds may even be necessary in some national parks. In Mediterranean Europe gathering fungi for the table is a popular family pastime and, although restrictions are enforced in reserves, collecting is generally not frowned upon.

It is very important that visitors to any country wishing to collect and study its fungi always ascertain what rules and legislation apply to the country – and sometimes even to a particular locality, whether it is a public space or under the control of a landowner.

The effects of deforestation

Although clear felling of forests and woods can have a devastating effect on fungal communities, there are important ways of mitigating the damage. It is beneficial to leave patches of natural vegetation within the felled and replanted areas, even if the latter are predominantly introduced tree species. These patches act as corridors facilitating the short-range dispersal of spores between one site and another, aided by small eddies of air, or by becoming attached to bird feathers or animal fur or in their droppings. Worms and insects are important agents in redistributing spores that find themselves on the soil surface or beneath it. Some spores are dispersed over much longer distances in upwelling air currents. Fungi can even survive in plantations if native trees are present and, given the right opportunities, they can then become more widely distributed.

Small copses and windbreaks can be havens for larger fungi, and introducing 'set-aside' as part of the European Union's Common Agricultural Policy has encouraged the planting of trees on former agricultural land; this can only be beneficial. Also, recreational areas, such as golf courses and parks, offer a 'half-way house' between natural communities and urban sprawl, and many vulnerable species can be found in these semi-protected areas.

In other parts of the world, including the tropics, forest and farm management is drastically changing to fit the demands of modern society. Now is the time to design and implement management strategies that include fungi, especially those involving the choice of tree species – a paramount factor in any new projects as far as fungi are concerned. For instance, larger fungi in the UK are more frequently associated with beech, birch, oak and their relatives and Scots pine, and these trees should play a major part of any planting programme, rather than species such as sycamore and rowan, which have a limited range of associated fungi. Lime trees, once thought to rely on microfungi for their mycorrhizas have, over the last twenty five years, also been shown to be ectomycorrhizal, supporting a diverse range of larger fungi. All

vascular plants have associated microfungi and it is necessary to strive for as varied a mosaic of plants as possible to ensure the survival of a range of species.

Although plantations are not considered of great biological interest by many naturalists, there is evidence that they too can support as diverse an assortment of fungi as native woodlands. Indeed, native species of fungi can take up residence in and on introduced non-native tree species, and in this

BELOW **The larch slippery jack,** *Suillus elegans* **associated with planted larches (***Larix* **sp.)**

respect are unlike many other native organisms, such as insects. Indeed, the non-native trees can act as a source of future inoculum for fungal associates of native plants. It is nevertheless true that many plantations contain common and widespread species of fungi, so perhaps it is the quality of the biodiversity that should be judged and not simply numbers.

Surprisingly, some of our most familiar and frequent mushrooms were originally introduced from other countries; a good example is the larch slippery Jack, bought in with European larch at the beginning of the 19th century. There is also some evidence suggesting that the closer a plantation is to native or semi-native woodland, the greater diversity of fungal species it contains.

Many British woodlands have one major disadvantage as places for fungi to thrive that sets them apart from woodlands in other parts of the world – except in developing countries, where firewood for domestic use and charcoal preparation is at a premium. This is that they are frequently too neat and tidy, having been cleared of their fallen trees, branches, woody fragments and tree stumps. All these act as a food source for larger fungi, as well as for countless microfungi and invertebrates, and it is essential that enough of them are left in place if the full diversity of fungi is to survive. Some wood-rotting fungi, having persisted unseen in a latent form high in the canopy, within the bark of the host tree, can fruit very quickly when a branch falls to the ground. Other fungi require a much longer gestation period to colonize an area, and are found only in very old woods;

several species are known only from primeval forests. To appreciate the range of fungi to be found in a particular locality it is necessary to undertake methodical collecting over a long period of time.

Creating a mosaic

The range of fungal species found in any country today results from its biological history and the effects of human activities, and new species are still being added on a regular basis. It is necessary to pay attention to the rare or infrequent species and also to accept that even introduced species need some protection. As fungi have been so poorly studied, it is difficult to assess the damage they experience as a result of changes in specialized habitats – except for the most obvious examples, such as the total destruction of a woodland or pasture. In the end, it may prove impossible to manage a habitat to benefit all its fungi, as each of the components, from decomposers of wood to ectomycorrhizal species, requires different conditions in which to fruit. A mosaic of vegetational types is therefore necessary, and theories of island distribution might have to be considered in any general management plan.

Fungal hotspots

There are undoubtedly places with high concentrations of fungal fruiting, but they do not necessarily reflect the distribution of rare species; in fact, mycologists do not understand the true biological significance of such sites. Perhaps they reflect the richness of the area in historic times and indicate what might have been present before early human interference. If this is the case, they serve as important refugia (areas that have remained unaffected by environmental changes occurring around them, and thus still contain the original wildlife of the region).

Although the fruiting of specific fungi was not always appreciated by naturalists in Europe, many field workers almost instinctively knew which were the rich areas where they could find a wide range of fungi. This treasure chest of field experience, accumulated over many years, is now beginning to be unlocked and in Britain, through the conservation organization Plantlife, a national scheme of fungal hotspots and how to identify them has been devised. As we gain more knowledge, it will probably be necessary to modify the scheme. This exercise will help planners, nature reserve wardens and other officials in the task of everyday management.

The scheme will identify areas that should be recommended for some kind of conservation of their fungal diversity, and is not only based on the existence of rarities at a site. Many of the fungi involved interconnect with each other as they are characteristic of the same special habitat, and fruit under very similar conditions. Thus members of several different genera of tooth fungi are known to fruit close to one another at scattered locations, in apparently suitable locations. Parallels must exist elsewhere.

Hopefully, negotiations to apply similar conservation criteria to other European countries will be successful in the near future, and then worldwide, with the scheme as part of an overall global strategy.

Fungi and government legislation

Each country addresses the issue of the legal protection of fungi in different ways, within both their civil and criminal legislation. However, worldwide, fungi appear in three categories. The first is within civil law – for instance, where a necrotrophic mushroom or bracket fungus causes the demise of a tree which may fall or fragment, and damage persons and/or property, often involving insurance claims and litigation. The second is within the criminal law – as with the illegal use of a specified group of larger fungi as recreational drugs, the adulteration of drugs and foods for profit, breaches of quarantine law, or the crimes of trespass and theft that may be committed by those selling wild mushrooms. Poisonous fungi have even been involved in murders, including those of certain Roman emperors such as Claudius – though mostly in the realm of fiction! The third legal category is that involving conservation legislation, touched on above and below.

In addition to the laws relating to conservation policy, there are two other important legal aspects involving larger fungi. The first is the illegal removal of specimens, for instance by members of a biological expedition visiting a host country. There are international agreements on the movement of plants, and fungi are included therein. Several developing countries are rightly sensitive to exploitation of their resources and impose regulations aimed at ensuring that any collected specimens or cultures of fungal hyphae that have been made must stay in the country of origin, with only duplicate specimens and cultures being brought back to the country sponsoring the collecting trip.

The second important aspect of international law is the breaking of plant quarantine laws. The latter is a serious offence as it could threaten the country's economy or its food production. Earlier in this book examples are given showing just how easy it has been in historic times for fungal spores or other vegetative parts to be brought by accident into a new country; hence, the strict implementation of quarantine law is paramount e.g. since its appearance in Europe hundreds of years ago, the dry-rot fungus has done incalculable damage. Different countries have different requirements for quarantine.

Government policy

In northern Europe, government policy on the protection of fungi today, in contrast to the situation ten or so years ago, is an important component of conservation policy, especially as the important role fungi play in all ecosystems becomes more widely accepted. Because of the availability of funds, some of the Western nations are leading the world in aspects of fungal conservation and have been able to monitor the success of these activities through such organizations as the IUCN.

At present, the provisional Red Data List for the British Isles contains 583 species, but now that more critical work is being undertaken in the field over the last few years this figure may be reduced as a result of more rigorous selection. Already, 27 listed species have been afforded protection by the UK government as there is concern about their

future. Many of these are either ectomycorrhizal, and therefore linked to a specific host, or found only on veteran trees or a small range of species.

Perhaps the best studied fungal community is that of the wax-caps occurring in unimproved grasslands, which have been studied on a Europe-wide basis. It has been suggested that if 17 species of wax-cap are known from a grassland site or if 11 species are found on a single visit, the locality should be considered of national importance. Scotland is very rich in wax-cap pastures, many of which have been found to meet the above criteria. Apparently, some modification of this scheme will be necessary for different countries or vegetational zones. This precision has been made possible only by widespread fieldwork enriching the compilation of species-lists from many previously unstudied areas.

Although more than 500 species of fungi currently attract British conservation interest, little is known of their biology; elsewhere in the world, the picture is the same. Not only is more research required but also a better basic understanding, as early studies suggest that the fungi are part of a complex interaction between the soil and its fauna and flora, both of which can be affected by even small changes in climate. The creation and application of conservation legislation for fungi presents a real challenge for the future.

Although they may be invisible to us for much of their lives, we owe the members of the fifth kingdom a great debt for maintaining the health of the world's ecosystems and, including the microfungi, for supplying us with a whole range of most important commodities.

Glossary

agaric soft, putrescent fungi with an umbrella-shape consisting of a stem surmounted by a cap under which are borne plates called gills; general term drawing together mushrooms and toadstools

agaricoid a fungus resembling those putrescent fungi with gills underneath a cap surmounting a stem

annulus a ring-like collar around the stem of an agaric; also called the ring

arbuscular mycorrhiza endomycorrhiza formed by members of the *Endogonales*, a group once placed in the phycomycetes

ascomycete(s) fungus which produces its spores in a sac (ascus), scientific term Ascomycota

ascus the sac in which the spores are formed in the Ascomycota

Attamyces genus of fungus associated with leaf-cutting or attine ants of the genus *Atta*

basidiomycete(s) fungus which bears its spores at the apex of a cell (basidium), scientific term Basidiomycota

basidium(a) the cell on the summit of which the spores are borne in the Basidiomycota

bioluminescence a phosphorescent glow which can be seen in the dark, produced by certain organisms including some fungi

biotroph process by which nutrients are obtained through a relationship with another living organism

bracket fungi refers to a range of larger fungi with pores (polypores) which form bracket shaped or shell-shaped fruiting bodies on woody materials: Basidomycota

cage fungus those members of the phalloid fungi which form a cage-like fruiting body covered in a smelly mucus: Basidomycota

clavarioid fungus with the shape of a club or coral, like a member of the genus *Clavaria*: Basidiomycota

corticoid a resupinate, crust-like fruiting body which possesses a smooth or slightly irregular surface as in members of the family Corticiaceae; Basidiomycota

cyanobacterium a bacterium that contains blue-green photosynthetic pigment and which can form a symbiotic relationship with particular fungi to form lichenized fungi

cystidium(ia) a differentiated end-cell on or in any structure of a basidiomycete fruiting body

decomposer fungus which breaks down dead material and litter to form simpler compounds; alternate name for saprotroph

deuteromycete a group of fungi brought together because they do not produce a sexual stage, or an asexual stage has not so far been discovered (alternative term for Fungi Imperfecti)

earthball member of the genus *Scleroderma*, the Latin name referring to the thick outer skin which encloses the spore-mass: Basidiomycota

earth star member of the genus *Geastrum*, so-called because the outer skin turns back at maturity to look like a star, the spore mass being found within a sac in the middle: Basidiomycota

ectomycorrhiza where the fungus forms a sheath or envelope around the root of the vascular plant which assists in the transfer of nutrients

endomycorrhiza where the fungus is inside the cells of the host vascular plant, with connections assisting in the transfer of nutrients with the outside soil

endophyte fungus which occurs inside the tissue of plants without showing signs of its presence

enzymatic special proteins which catalyse the chemical breakdown of organic compounds

family term used for groupings of similar genera

flagella(ae) term applied to specialized structures which assist in the movement of spores

gasteromycete group of fungi in which the spores are enclosed within a fruiting structure. At one time erroneously thought to be a natural grouping and given the name *Gasteromycetales* (also called stomach fungi): Basidiomycota

genus(era) term used for groupings of similar species

honey fungus a name applied to members of the genus *Armillaria*, generally *A. gallica*: Basidiomycota

humicolous fungus which grows directly on soil intermixed with derivatives from the breakdown of woody material, leaves etc. Such a fungus is sometimes called a humicole

hydnoid those fungi that produce their spores on teeth-like projections; they are both stalked or resupinate species; members of the genus *Hydnum*: Basidiomycota

hymenium the tissue in which the basidia or asci are arranged

hypha(ae) the filament which emerges from the germinating spore and forms the basis of further growth

jelly fungi name applied to two large groups of fungi with septate basidia and whose fruiting body is highly gelatinized: Basidiomycota

K-fungi highly competitive fungi

larger fungi general term for mushrooms, toadstools and their relatives (Basidiomycota) in addition to the morels and a few larger elf-cups (Ascomycota), i.e. those species collected by naturalists on a foray (alternative name for macromycetes)

lichenized fungi fungi which are in an intimate association with algae; commonly simplified to lichen

metabolite chemicals manufactured during the growth of an organism

morel member of the genus *Morchella*: Ascomycota

mushroom general term applied to an edible or non-toxic agaric

mycelium aggregation of hyphae

mycorrhiza term indicating the state of a fungus when it is in a mutualistic relationship with a vascular plant

mycorrhizal fungus a fungus in a mycorrhizal relationship

mycota general term for all the fungi that might be found in a single locality

myxomycete group of fungi (alternative name for slime moulds) now because of their mode of feeding considered more closely related to the amoebae

necrotroph term applied to fungi which colonize living tissue bringing about its death and then living on the remains

parasite term applied to a fungus which obtains its nutrients from the cells of living organisms whilst in some cases maintaining the living, albeit debilitated, host

phalloid fungus with the features of a stinkhorn – member of the genus Phallus, or sometimes applied to the whole of the order Phallales: Basidomycota

photosynthesis process by which green plants manufacture food material from water, carbon dioxide and sunlight

phycomycetes a general term, at one time erroneously considered a natural group of filamentous fungi which lacked asci and basidia and had some characters in common with algae

polypore generally tough, woody or leathery fungi with the spores borne within tubes; fruit-body is either stalked or resupinate in different species

poroid fungus with the sexual tissue distributed inside long or shallow tubes as in the genus *Polyporus*: Basidiomycota

puffball name for a group of stomach fungi where the spore mass is contained within an outer thin membrane and therefore resembling a sac: Basidiomycota

resupinate where the sexual tissue in the basidiomycete resembles a sheet or crust without cap or stem; may have pores, teeth, wrinkles or be quite smooth

ring see annulus

ruderal fungus fungus which takes quick advantage of the availability of a nutrient source

saprotroph fungus which lives on dead material

sheathing mycorrhiza where the fungus forms a sheath or envelope around a tree root to assist in transfer of nutrients, also called ectomycorrhiza

species the basic unit into which scientists divide the living world; the members of the same species can interbreed with one another, but not with members of another species

spore a general term for the dispersal propagule (diaspore) of a fungus of which there are many types

spore-print the pattern obtained on a surface from the ejected spores of a basidiomycete

stromatolite fossilized mixtures of primitive organisms from shallow water, generally cyanobacteria

Termitomyces a genus of agarics which are intimately associated with termites: Basidiomycota

toothed fungus fungi with spines replacing the gills of an agaric or pores of a bracket

vegetative asexual

volva the cup shaped structure found at the base of the stem in certain agarics

yeast fungi free living, single-celled fungi or stages in the life-cycle of a range of fungi, especially members of the Ascomycota. Commonly abbreviated to yeasts and often refers to those used in bread, beer and wine making

Index

Further Information

Websites

British Mycological Society
http://www.britmycolsoc.org.uk
Information about the society, calendar of meetings, publications and related information. Links to *Mycological Research* and *Mycologist*, and *Field Mycology*.

Fungimap
www.rbg.vic.gov.au/fungimap/
Aims to record the occurrence and distribution of Australia's mycoflora using a network of volunteers across Australia. Includes identification tips for fungi, information about edible and poisonous fungi, images of key species and 'strange' facts.

Images of fungi
http://www.in2.dk/fungi/
Contains links to over 1600 images of fungi on the internet.

Mycological Group, Royal Botanic Gardens, Kew
http://www.rbgkew.org.uk/scihort/mycolexp.html
Provides an overview of current research and a general introduction to fungi, with information on the mycological specimen collection.

Mycological Society of America
http://www.msafungi.org/
Dedicated to advancing the science of mycology with comprehensive links.

MykoWeb
http://www.mykoweb.com/
A collection of web pages with links to other internet resources and a directory of mycological societies in North America.

New Zealand
http://nzfungi.landcareresearch.co.nz/html/mycology.asp?ID=84-HMB-80
Access to "a wealth of information on fungi in New Zealand". One database contains images, species and group descriptions and some identification keys.

University of Sydney
http://bugs.bio.usyd.edu.au/Mycology/default.htm
Offers an online learning package, aimed at an undergraduate audience. Each section has images, descriptions and question and answer modules.

Virtual Library for Mycology
http://mycology.cornell.edu/
An extensive directory of links to mycological resources including directories of mycologists and labs, collections, molecular genetics, guides and teaching.

Further Reading

Ainsworth and Bisby's Dictionary of Fungi, P.M. Kirk, J.C. David and J.A. Staplers (eds.) CABI, Wallingford, 2001.

Children and Toxic Fungi, R.Watling. Royal Botanical Gardens, Edinburgh, 1995.

Co-evolution of Fungi with Plants and Animals, K.A.Pirozynski and D.L.Hawksworth (eds.) Academic Press, 1988.

Collins New Naturalist Series: Mushrooms and Toadstools, 2nd edn. Godfrey Cave Associates, London, 1989.

Fundamentals of Mycology, J.H.Burnett. E.Arnold, London, 1968.

Fungal Conservation: Issues & Solutions, D.Moore, M.M. Nauta, S.E.Evans and M.Rotheroe (eds.) Cambridge University Press, 2001.

Fungal Ecology, N.J.Dix and J.Webster. Chapman & Hall, London, 1995.

Fungi: Folklore,Fiction and Fact, W.P.K.Findlay. Kingprint, Richmond, 1982.

Fungi, Man and his Environment, R.C.Cooke. Longman, London, 1980.

Fungi: Naturally Scottish, R.Watling and S.Ward. Scottish National Heritage, Perth, Scotland, 2003.

Fungi of Southern Australia, N.L.Bougher and K.Syme, University of Western Australia, 1998.

Growing Gourmet and Medicinal Mushrooms, P.Stamets. Mycomedia, Olympia, Washington, 1993.

How the Mushroom got its Spots: An Explorers Guide to Fungi, S.Assinder and G.Rutter. British Mycological Society, 2003.

How to Identify Edible Mushrooms, P.Harding, T.Lyon and G.Tomblin, Harper Collins, 1996.

Insect-Fungus Interactions, N.Wilding, N.M.Collins, P.M.Hammond and J.F.Weber (eds.) Academic Press, 1989.

Introduction to Fungi, J.Webster. Cambridge University Press, 1970.

Introductory Mycology, C.J.Alexopoulos, C.W.Mims and M.Blackwell. Wiley New York, 1996.

Lichens, W. Purvis, The Natural History Museum, London, 2001.

Managing yourland with fungi in mind (A5 leaflet). The Fungus Conservation Forum 2001.

Mushrooms and Other Fungi of Great Britain and Europe, R.Phillips. Pan Publications, London, 1981.

Mushrooms and Toadstools of Britain and Europe, R.Courtecuissse and B.Duhem. Harper Collins, 1995.

Mushrooms of Northeastern North America, A.E.Bessette, A.R. Bessette and D.W.Fischer. Syracuse University Press, New York, 1997.

Mycorrhizas in Ecosystems, D.J.Read, D.H.Lewis, A.H.Fitter and I.J.Alexander (eds.) CABI, Wallingford, 1992.

Slayers,Saviors, Servants and Sex, D..Moore. Springer Verlag, New York, 2000.

Picture credits

Front cover Photodisc; inset (US only) Roy Watling; back cover and title page Roy Watling; p.5 Richard Revels; p.7 Roy Watling; p.8 left David Hibbett & *American Journal of Botany*; p.8 right, p.9 Roy Watling; p.10 RBGE; p.11 Richard Revels; p.12 Mike Eaton/© NHM; p.13 left Dr R.F.O. Kemp; p.13 right Dr Martyn Ainsworth; p.14 Mike Eaton/© NHM; p.15 Dr Martyn Ainsworth; p.16 top Roy Watling; p.16 bottom © N.W. Legon; p.19 © Frances Fawcett/*Systematics Agenda 2000 Technical Report*; p.21 left Dr Maurice Moss FLS; p.21 right Mike Eaton/© NHM; p.22 © N.W. Legon; pp.23-25 Roy Watling; p.26 top © Greg Basco/www.deepgreenphotography.com; p.26 bottom Roy Watling; p.27 top left Gordon Dickson; p.27 top right and bottom Prof John Hedger; p.28 Richard Revels; p.29 © N.W. Legon; p.30 George Beccaloni; p.31 top and bottom © N.W. Legon; p.32 left and bottom © Sidney J. Clarke FRPS; p.32 right © N.W. Legon; p.33 top Gordon Dickson; p.33 bottom, p.34 © N.W. Legon; p.35 Dr Martyn Ainsworth; p.36 left and right Gordon Dickson; p.37 top Dr Martyn Ainsworth; p.37 bottom © Sidney J. Clarke FRPS; p.38 Roy Watling; p.39 Dr Martyn Ainsworth; p.40 Roy Watling; p.41 top and bottom © N.W. Legon; p.42 Gordon Dickson; p.43 Mike Eaton/© NHM; p.44 Gordon Dickson; p.46 Dr Martyn Ainsworth; p.47 Prof OK Miller; p.48 Gordon Dickson; p.50 top and bottom George Beccaloni; p.51 Anthony Bannister/NHPA; p.52 Roy Watling; pp.53-54 Gordon Dickson; p.55 top Geoffrey Kibby; p.55 bottom, pp.56-57 Roy Watling; p.58 Yves Lanceau/NHPA; p.59 © Simon Moore/Hampshire County Museums Service; pp.60-61 Dr Maurice Moss FLS; p.62 Mike Eaton/© NHM; p.63 © N.W. Legon; p.64 Mike Eaton/© NHM; p.65 William Tait; p.66 left Roy Watling; p.66 right Dr Martyn Ainsworth; p.67 Roy Watling; p.68 Oakleaf European Ltd; p.69 British Wood Preserving and Damp-proofing Association; p.70 © N.W. Legon; p.71 © Sidney J. Clarke FRPS; p.73 top Richard Revels; p.73 bottom Gordon Dickson; p.74 Roy Watling; p.75 Gordon Dickson; p.76 top © N.W. Legon; p.76 bottom, p.77 Gordon Dickson; p.78 top Roy Watling; p.78 bottom Gordon Dickson; p.80 William Tait; p.84 Gordon Dickson; p.86 Scottish Wild Mushroom Forum/Scottish Executive; p.88 Gordon Dickson.